YOUR NOSE KNOWS

YOUR NOSE KNOWS
Reviews

PERFUMERY AND
Essential Oil Record

Vol 52 Number 12 December, 1961

(Published Monthly G. & M. Press Ltd.)

33-34 Chiswell Street,

London, EC. Great Britain

"...A new hypothesis of human motivations and behaviour is put forward and all who are interested in the subject of smell should read this book, since in it the effects of odour on hu man beings from a behaviour point of view are set out in great detail. It is recommended as a mind stimulant even though the reader may be at cross purposes with what has been written."

Biological Abstracts

June, 1961

University of Pennsylvania,

3815 Walnut Street,

Philadelphia 4, PA

"Man's sense of smell is acted upon by both 'objective and subjective' odors. Information so derived is accepted subconsciously and plays a large role in man's behavior. Automatic actions of the body are primarily responses to odors..."

Eastern Pharmacist

October, 1962

New Delhi-16

India

"...The reader is bound to be enveloped with the fervour of the author once he has started going through the pages of the book. There are undoubtedly large amounts of truth in what the author says on smell and its ramifications. The book is a mind stimulant."

Philosophy and the Arts
A LITERARY AND PHILOSOPHICAL REVIEW

Number 13, March 22, 1983

P.O. Box 431, Jerome Avenue Station,

Bronx, N.Y. 10468

(Reviewed by Daniel Manesse)

"Here is a remarkable study of the least esteemed of our five senses—the sense of smell. This study has been a lifelong project for its author. One of [his] conclusions is that a real division and conflict exists between our primitive animal brain which is dominated by the sense of smell and our lately acquired consciousness which is under eye and ear control... Nevertheless, the smell apparatus has continued to act both subtly and powerfully upon our behavior. Chronologically, it was the first of our sense organs to be developed. And it is the only one which has a direct connection to the brain.... Long ago, Sigmund Freud noted that the female odor was openly recognized as a powerful sexual stimulant in some cultures. Although Millen is not favorably disposed towards psychoanalysis, he appears to have made a valuable contribution to his field. His study of the sense of smell has thrown considerable light upon the workings of the id—that part of the personality which Freud identified as the source of our unconscious drives. It is to be hoped that "Your Nose Knows: A Study of the Sense of Smell" will at long last receive the recognition it deserves."

OFFICIAL NEWS PUBLICATION
PHARMACEUTICAL SOCIETY OF PAKISTAN
THE CHEMISTS REVIEW
Vol. 7, March-April 1963, No. 3-4

17 Gulshan Terrace,

Bunder Road, Karachi, Pakistan

"The book is more fiction than pure science. None the less it develops more the common border of science and fiction…and brings out unbelievable facts discovered by the author."

THE *Pharmacy* NEWS
MONTHLY

Incorporating "Modem Medical Practice"

VOL.12 March 1962 NO.3
LUDHIANA

EAST INDIA PHARACEUTICAL WORKS LTD.

CALCUTTA-26

India

"Everybody has a nose, but few know the mysteries of the main function of the nose-the sense of smell. In this book the able author James Knox Millen has ably explained the mysteries of nose and contains a philosophical study of the sense of the smell, and really, it will give you a new introduction with your nose."

Ted Emery, Busch Chair,
THE WHARTON SCHOOL
of the
UNIVERSITY OF PENNSYLVANIA

(Letter to the Author, October24, 1982)

BUSCH CENTER, Vance Hall CS

3733 Spruce Street,

Philadelphia, PA 19104

..."I took up the Busch Chair here in July. I posted on separately, from Australia, 10-30 of what I regarded as my invaluable, irreplaceable books. Less than an hour ago I was bemoaning their fate (and mine) to my wife. I mentioned your book as an example of what I would miss. She told me about the notice in N.Y.T. I have never had any doubt about the value of your insight. Anyway [since I can again acquire the book], all is not lost to me."

YOUR NOSE KNOWS

■

A Study of
THE SENSE OF SMELL

by
JAMES KNOX MILLEN

Authors Choice Press

San Jose New York Lincoln Shanghai

YOUR NOSE KNOWS

Authors Choice Press
an imprint of iUniverse.com, Inc.

For information address:
iUniverse.com, Inc.
5220 S 16th, Ste. 200
Lincoln, NE 68512
www.iuniverse.com

Originally published by James K. Millen/Cunningham Press

ISBN: 0-595-01208-6

Printed in the United States of America

A FOREWORD

IN THIS BOOK I offer an hypothesis—a new theory of human motivations and behavior. It is based on three assumptions: that one and one (of similar things) make two; that (failing physical proof to the contrary), what was, *is;* that an animal (man) acts like an animal in his most animal-like actions.

You will be reluctant to accept it. Understandably so, if you are a professor or practitioner of Psychology. Neither would witch-doctors enjoy the realization that mumbo-jumbo is wrong, and true medicine-men should use the A.M.A.'s stock-in-trade: antibiotics and aspirin. No man likes to see the ground cut from under his feet; no man wants to discover that nearly all he has taught, and written, and halfway believed in is untrue.

But even if you are the usual reader, and this is no matter of life-and-livelihood to you, you are not going to accept this hypothesis. Rather, you will find yourself violently unwilling to believe it, angrily anxious to prove it false.

Now, this is a curious thing. Theories are as many as the sands of the sea. Normally one considers them, accepts them in part or in whole, or rejects them. Calmly, equably—never with violence like this. Forget everything else, and consider this surprising anger.

In this book I speak of a taboo. Ask yourself in all honesty if the strange violence of your rejection is not in some sort confirmation of the taboo's existence.

The matter is important. Because it explains why this hypothesis—a thing seemingly as plain as the nose on your face—has never before been offered.

You will not quickly believe this book—but neither will you quickly forget it. And whenever you encounter the fumbling, faltering attempts-at-explanation of Personality, and Likes-and-Dislikes, and Dreams, and other enigmas present-day Psychology cannot explain, you are going to remember that there is a theory that explains them, simply, straightforwardly, adequately.

The schoolmen cannot ignore this hypothesis forever. Sooner or later, they must accept it, or disprove it: it cannot work, or it does not work. And since it concerns itself with, and depends upon the workings of the human body, the disproof must be in and of such workings—not (as over and over I have been told, as if this settled the question) that the thing must be untrue because Sigmund Freud or Till Eulenspiegel or John Doe said something to the contrary. The days of the Scholasticists, when no science was acceptable unless Aristotle said it first, are supposedly 600 years past.

Without question there are errors many and small in this book, and they will be seized on gleefully as disproof. As to that, I would point out that the early cartographers made mistakes a-plenty in mapping the curves of the river's channel, but these inaccuracies in no way disproved the Mississippi's far-reaching presence.

So frequently that I have come to expect them, certain criticisms— which the authors seem to consider damning—have been made about this book:

"Why, if your hypothesis is so simple, so reasonable, hasn't it been offered before."

Well, only the writers of high-school textbooks adopt the attitude that we know everything. Our grandchildren will be discovering things our grandparents should have thought of.

There is a specific corollary to the above question: *"Why haven't the professors of Psychology discovered it? Why was it left for you, a layman . . .?"* and so forth.

This one, I am going to answer specifically. Too much, too-formal study seems fatal to original thought. The mountainous research laboratories of the country, staffed to the top with Ph.D.'s, labor and bring forth microscopic mice. The do-nothingnesses of the Ford Foundation, the speed with which we are falling behind Russia in the current space-race, are other instances of what happens to creative progress when the professors take over.

And now, for that "layman" sneer, that background-of-authority insistence: five thousand years ago the Egyptian priests—the professors of the day—proclaimed knowledge, and the furtherance of it, an esoteric thing, an arcanum reserved for their own initiates; and continually, through the centuries since, their successors have asserted and/or assumed the same authority. Sometimes it has been granted them, and it might be mentioned in passing that these were the years when the world stood still—the so-called Dark Ages.

Even in modern times, and often, schoolmen show the same biased belief: that research should be a closed shop, with the union card a graduate degree—and there is the same jealousy of and rancor toward outsiders. Edison the exoteric was sneered at; Wallace, the unknown, had his evolutionary revolutionary paper on Natural Selection held up by the Royal Society until Darwin, one of the elect, could prepare his paper, supposedly discovering the same thing. (And which one to this day, gets his name on, and credit for, the discovery?) Against all reason, the Smithsonian Museum insisted that its own Professor Langley, and not two bicycle mechanics, invented the airplane. . . .

There is a pretty story, many times textbook-told: of the self-effacing little country priest who crossed wrinkled peas and smooth, green peas and yellow. From these experiments, he drew certain conclusions which in all modesty he printed in his parish magazine.

The world knew nothing about it; the little priest died and was forgotten. Twelve years later, quite by accident, a scientist came on a copy of this magazine; realized the significance of the priest's experiments, and gave his tremendous results to the outside world.

So one more book was added to the scientific Bible; too late, the little priest received his rightful acclaim; his hesitant theory became Mendel's Law, the very basis of the modern science of Heredity.

A pretty story. The trouble is: it simply isn't true. Gregor Mendel was no more modest than the next man, and had as little desire to blush unseen. He bombarded the scientific societies of Europe with his findings; was greeted with cold contempt ("Another outsider with a crazy idea! . . . How many degrees has *he* got!"); and published his theory in the parish magazine only because he couldn't get it published anywhere else. The one bit of truth in the standard story is that he died and was forgotten.

Safely forgotten, one eminent savant thought. Eager for fame, he published the dead priest's findings as his own invention. Unfortunately, the same easy path-to-the-stars occurred to two other professors, simultaneously. So, in 1896, *three* noted scientists announced the same discovery, all at the same time. The academic welkin rang with accusations and counter-charges. Each professor and his partisans called the other two thieves.

And then, a fourth pedant, jealous of all three, brought out a copy of the parish magazine, and proved that everything the others had said about each other was true. They'd all been filching from dead Gregor.

By now, the situation was what could only be termed confused. No one else dared claim the theory; somebody had to be given credit for it; Gregor Mendel, the outsider, received his rightful due.

All men tend to exalt their professions. The schoolmen's belief that learning should be limited to a closed corporation is understandable and would be harmless were it not (the Big Lie technique!) that the world to an extent believes it.

I say to you that the graveyards are full of degree-laden scholars who have contributed not one tittle to the advancement of knowledge;

I say to you that the world has been bettered in a million ways by the work of outsiders whom the schools scorned.

With all the sincerity I possess, I say to you that it is the idea, and not the author thereof; that it is the thought, and not the thinker, that matters.

Research on this book was completed in 1932. Under its title: YOUR NOSE KNOWS, and in a rougher form, it was that year submitted to a dozen publishers. In 1937, at an editor's request, I rewrote it, polishing it as much as I could. But I subtracted nothing; I changed no thought, no argument; I added just one thing: the unimportant bit about green eggs.

To all intents, this book is the 1932 version. Except for the present preface and three footnotes, it is exactly as it was offered in 1937.

There was a temptation to rewrite it. My first manuscript was dogmatic; this one errs on the apologetic side; perhaps I could find a middle ground. I doubt if I would keep the sweetness-and-light ending; an old man is not that hopeful of anything. I could shorten the discussion of Behaviorism; it is not so important, now. I could list solid citizens who feel as I do about Psychoanalysis, since all the world is attacking it, today.

But—if my hypothesis was true in 1932, it should be true, now. And, it seems to me, revelation of how well it holds up under the buffeting of 28 years and whatever new wisdom these years have brought should be an excellent test of its validity. And so, I put it to this test, unchanged.

But I am not being so greatly daring, after all. There has been miniscule progress. For example: in this book I go over the experiments on instinct described in a General Psychology used at Yale in 1931. Last week, as a check, I examined the material on the same subject in a textbook current at the University of California today. Exactly the same experiments are listed, the same conclusions reached. Last year the Reader's Digest published an article on the Sense of Smell. The latest information, supposedly. Every statement in it can be found in the first few chapters of my own 1932 manuscript, seen by scores of persons. And there is no question of plagiarism. The Digest author went to the same sources I did: books and articles half a century and more old— and he did so because he could find nothing newer, because there was, and is, nothing newer.

CONTENTS

YOUR NOSE KNOWS

CHAPTER ONE

VERY LITTLE is known about the sense of smell.

A strangely static very little. The textbooks dismiss it with a casual page or two; the encyclopedias give it less; the bibliography for reference turns out to be a single small book, devoted mostly to catfishes. Beyond this, there's a blank wall, an apparent dead end to research.

Nobody cares, particularly. Everyone knows that man's smell-sense is a weak, degenerate thing. Science has larger problems; civilization rocks along . . . so what?

This! Logically we should be quite upset over the continued dark spot in our firmament of wisdom. For weakly degenerate as it may or may not be—the sense of smell can stupefy or revive you, sicken or heal you. It can make you glad, make you sad, make you a John-o'-Bedlam madman. It has incredible control over emotions, memories, associations, moods.

It is to all intents *two* of the five senses, true taste being but a subsidiary part of it. It is one of the four roads by which life outside comes to us, one of the four means by which we know our world, one of the four reasons why we behave as we do. Fuller knowledge of it would indeed be worth while.

So then should this book be of value. For here, fairly written, are facts, and ideas these put-together facts have fathered, and surprising but imperative conclusions therefrom.

What your dreams mean, for instance, and something about ghosts and haunted houses, and why you like some people and dislike others, and why you go crazy, and fall in love, and . . . Various things.

Together now we—you and I—begin the study of Osmics, the science-that-is-not. the science of the sense of smell.

The very little that is known of it follows:

Smells are, technically, odors—but what are odors? Gaseous particles thrown off by the olfactory substance, say Parker* and the majority. Other scientists claim that odors are vibrations, simply.

---

* Parker: *Smell Taste and the Allied Senses in Vertebrates.*

In this book the first—the so-called chemical hypothesis—is accepted, because opinion favors it preponderantly. However, the question as to which theory is actually correct reveals this curious situation: both guesses can be proved wrong. Smell-reactions *must* be caused by one of the two—and yet the wave-advocates can prove that it is not a chemical sense, their opponents: that it is not a wave-sense. Havelock Ellis, seeking a middle ground, suggests that the particles come first, but the vibrations do the work.

The problem is not nearly so important as it seems. We've never got to the bottom of anything. Particles or waves? the same two opposing beliefs—the quantum- and wave-schools—war over the puzzle of what light is, but of light we've working knowledge, anyway. That gives us photographs and neon signs and the electric eye—plenty of useful things.

Ultimate knowledge is simply a goal to strive for. Working knowledge is enough. And fortunately, in the case of this wave-versus-particle conflict, it's beginning to settle itself. Physicists, so we are told, are coming to realize that the two forms of energy are ultimately the same.

Now for some difficult reading, too important to omit:

The structure of smell—the olfactory sense—is known. High in the nose are cavities (the olfactory clefts), one for each nostril. Inside these there is a total area of about one square inch, covered with a different sort of skin from the mucous membrane that lines the rest of the nose. This square inch (the olfactory epithelemium) is where the true sense of smell resides. It is surrounded by pigment cells (purpose unknown), is dark yellowish-brown in color, and for its size is better supplied with blood than any other part of the body. It is always covered with a thin fluid.

Through this watery covering and into the olfactory clefts, tiny hairlets protrude a very little. These, combining in clumps of six or eight, form the olfactory cells.

These in turn go through a spongy sort of bone (the ethmoid bone) without break or connection of any kind, and terminate in a leek-shaped organ inside the skull (the olfactory bulb). Here they end around other nerve cells (the mitral cells), which act as relay stations.

Nerve-endings from the mitral cells move in a thick, cable-like mass (the olfactory tract) and enter the hemispheres of the brain. The connection occurs in almost the exact center of the skull, straight back from the nose.

The olfactory cells are true nerves, and the little hairlets are true nerve-endings. The olfactory nerve, so-called, is really a nervous area, containing all these olfactory cells, which are the true carriers of sensation.

In considering the sense-organ, one fact is immediately apparent: the utter simplicity of the mechanism involved. In smell, this is absolutely all that occurs: a stimulus strikes a nerve-ending, causing a chemical reaction. This initiates a nerve-impulse which, by one connection, moves along another nerve into the brain.

The house of smell—the olfactory area of the nose—has a front and a back door, and both are useful. Outside stimuli, odors, enter through the nostrils. Funnel-shaped openings (the choana) in the rear of the mouth give entrance to smell-messages too: the delicate sensations of taste.

Coarse taste-sensations, we learn, come from the taste-buds along the tip, the sides, and the base of the tongue. These buds, which are, practically speaking, part of the sense of touch, react to four sorts of stimuli: sweet at the tip, salty at the sides and base, sour at the sides, and bitter at the base of the tongue. The taste-buds can distinguish broadly between these four sensations, and that is all. Flavor, aroma, *real* taste as we conceive it, come from back-door messages into the house of smell.

That is why, when the nose is stopped up with a cold: "Food has no taste."

As Dorsey says: "Without the sense of smell we cannot distinguish honey from molasses, black coffee from quinine, an apple from an onion. Tea and coffee-tasters are tea and coffee smellers."

Inside the nose there are pairs of ridges. These lateral folds are the conchae. Some of them (those attached to the maxillary bone) lie in the purely respiratory regions of the nose. Others, those from the spongy ethmoid bone, serve as olfactory surfaces.

Man has from three to five pairs of these (olfactory) ridges. Sixty per cent of adult humans, so Schaeffer says, have four pairs. Monkeys possess only three, and in mammals generally the number varies. A certain ant-eater tops the list with ten.

Because the keener-scented animals have more of these ridges, the number of pairs found is used as a rough-and-ready classification. Mammals that have the greater number are described as macrosmatic—keen-scented. Those species in which the number is four or less are considered to be less acute of smell (microsmatic). The whales and porpoises comprise a third and very small division. In these the smell-organ has almost completely degenerated, and they are supposedly devoid of the sense entirely (anosmatic).

Whenever the pundits write about smell, they make mention of a scientific curiosity found in the nose. This is the organ of Jacobson. The heavier books call it the vomero-nasal organ.

No one knows whether it is rudimentary or vestigial. It appears throughout the animal kingdom with little regard for the rules of evolution. In humans it shows in infancy as a short flattened tube, a millimeter long, opening by a tiny pore just inside the nose. Sometimes, in adults, it disappears.

Perhaps (here are the usual guesses) it is a subsidiary smell-organ. Perhaps it was formerly the middle one of an original group of three, and when the outside two came together, it lost out, so that only vestiges of it remain.

In the nose, as in all the other natural apertures of the body, there is a chemical sensitivity of the partly exposed mucous membrane. To the lay mind this would seem part of the sense of touch, stronger only because the cushioning cover-layer is thinner here, but we are told that this is not so. This sensitivity is off on its own as the common chemical sense, and "the common chemical sense has no relation to taste, smell or touch, and is only distantly suggestive of pain."

It has been argued that the organ of Jacobson has use as a receptor for this sense. Two crushing objections flatten this theory. One is: why aren't there other such organs in the other and also-sensitive body-openings. The second, even stronger, is this: so far as anyone can discover, the organ of Jacobson has no nerve-connections, no way of transmitting its reactions to the brain.

Ph. D.'s—a respectable bread-line of them—have gained their scholastic honors by original research on this organ. The sum of their contributions to the body of human knowledge is—we'll let Professor Parker say it: "Concerning the function of the organ of Jacobson almost nothing is known."

So we go on.

Endlessly scientists have busied themselves trying to classify the various odors of which we are conscious. Inasmuch as no two lists agree, and the classification of Linnaeus, who died two hundred years ago, is mentioned with as much or as little respect as the olfactory prism lately suggested by Henning, there seems no profit in considering them here.

Science has found that each individual differs in his ability to recognize smells. It has found that certain persons can smell some odors but are totally unable to detect others. To such it has given the name: "partial anosmics." It has involved itself in various bitter controversies as to the comparative smell-sensitiveness of men and women, old and young.

Zwaardemaker is a great man to students of the subject. He devised a machine (the olfactormeter) to measure the difference in the smelling-ability of individuals.

Somewhat to his amazement, he learned this: that the human nose is more accurate than any machine humanity has to test it with. For instance (the example most frequently quoted) the human nose can detect the garlic-y smell of mercaptan when there is only one part of mercaptan to four hundred and sixty billion parts of air.

Inside the nose, in man and in all the other vertebrates, there are other free nerve endings beside those of the olfactory cells. These are from the trigeminal or fifth nerve. Magendie, nearly a hundred years ago, claimed that this was the true nerve of smell, and that the so-called olfactory nerve was one whose function was wholly unknown. He made experiments and, he thought, proved his statement.

In the scientific war that followed he got nowhere. Valentin and Schiff offered experiments in rebuttal. Nature never abhorred a vacuum the way Science abhors the unknown—and Magendie was trying to create an unknown out of something Science had already tagged and catalogued. He was foredoomed to failure. The scientific decision that followed became dogma that nobody, since, has dared challenge.

And this is the decision: the trigeminal nerve is affected by irritants. The olfactory is the true nerve of smell. It is stimulated by delicate perfumes, aromas, and food odors.

Failure to recognize this fundamentally important distinction, Parker says, is responsible for much of the confusion in the realm of smell. In his own words: "Irritants or stimuli for the trigeminal nerve induce violent reflexes, respiratory and the like. True odors are in nature much milder, and seldom call forth strong responses."

Smell-exhaustion, under strong stimulation, it has been observed, is accomplished in a very few minutes. Certain odors seem to neutralize each other. Ammonia and acetic acid, for instance, both stimulate the nose, but when mixed they possess no odor, for they combine to produce the odorless ammonium acetate.

Valentin observed that when ether and balsam of Peru are smelled at the same time, each by one nostril, the odors are perceived not together but alternately. There is a similar rotation between camphor and oil of cloves.

Zwaardemaker prepared a list of odors which in some way serve to neutralize each other so that neither can be smelled. Cedarwood and rubber, for instance, in proportions of 2.75 to 14 cancel out.

This occurs only when the odors are applied at different nostrils. For this reason, it is claimed, the neutralization and conflict must have a central, inside-the-brain-itself origin.

Pavlow pointed out, in one of the most famous experiments ever made, the importance of true odors in exciting and in a way controlling the whole chain of digestive secretions.

I quote Parker again: "In man and other microsmatic forms much of the keenness of olfaction has disappeared, and yet . . . we are often surprised by the power of our olfactory organs. The organ of smell gives up old forms of stimulation and takes on new ones in a way that is almost incredible."

This surprise is common among scientists who have gone into the matter. Like the other half of a Siamese twin, it usually appears side by side, as above, with the statement that man's sense of smell is degenerate.

Historically, the smell-sense is the most ancient of all. Only the olfactory nerves are directly connected to the hemispheres of the brain. Hence it is inferred that the brain itself arose in connection with the sense of smell. The original brain was a smelling-organ.

Finally, the sense of smell is both an exteroceptor and an interoceptor. That is: it is an external sense-organ, located on the body surface, as are the other two distance-receptors, the eye and the ear. It receives stimuli from without the body, as they do.

But differing from them, it is an inside-the-body sense-organ, too. In itself, and through its auxiliary sense of taste, it affects internal and visceral functions. Suitable nerve-impulses from it cause mastication, swallowing, peristalsis, and glandular action. Unsuitable impulses inhibit these activities.

So ends our brief summation: all that either natural science or psychology has to say about the sense of smell. If there are abrupt transitions, dissociated statements, it is because each little fact that is known seems to have no connection with any other fact in this olfactory jumble.

It is possible to add a mass of information about the sense of smell in insects, birds and mammals, some of which is most interesting. Also, there is much material about perfumes—the perfumes which, as the lady authors say in the ladies' magazines to their lady readers, express the personality. Later we shall devote a chapter to each, but not now.

For, as it is the purpose of this book to disclose the almost incredible power of man's sense of smell, at the moment it seems needful to consider just how degenerate the human smell-sense actually is, and since this is scientific dogma, precisely what scientific grounds there are for the assertion.

Apparently the case for degeneracy rests on four legs, and only four. They are:

1. Common knowledge.

2. The vestigial appearance of the organ of Jacobson.

3. The fact that man has only four pairs of olfactory ridges (conchae), whereas some mammals have five, some hoofed animals eight, and the ant-eater ten.

4. The comparative size of the olfactory areas in man and lower animals.

And how well do the proofs hold up under scrutiny?

COMMON KNOWLEDGE: This, the man-in-the-street thinks, is conclusive enough. Scientifically, it is not. From a textbook I quote: "A science is *critical* in its attitude. It questions beliefs which are supported solely by hearsay evidence, superficial observation, or prejudicial opinion. Even the evidence of 'common sense' has often been criticized by science ... Psychological science in particular frequently conflicts with popular or common sense opinions ... (Though) they constitute so integral a part of our common stock of ideas that we tend to regard them as almost self-evident or axiomatic . . . the psychologist has not hesitated to question these generalizations."

Brave, heartening words! Even though, parenthetically, no psychologist has questioned the above generalization—at least we have scientific permission and tacit encouragement to do so.

THE VESTIGIAL APPEARANCE OF THE ORGAN OF JACOBSON: The organ of Jacobson, considered as a vestigial remnant, and offered as a proof of degeneracy in man's sense of smell meets a perfect analogy in the pineal gland, man's so-called "third eye." A like hypothesis explains this as a middle eye, lost sometime in the past after Nature learned the advantages of bilateral symmetry. Lizards, in which the organ of Jacobson is most highly differentiated, have also a less vestigial third eye. One survivor of old eras, the sphenodon, uses it to distinguish light from darkness.

Every man's head holds the little pineal gland, filled with its useless brain dust, but does anyone call human eyesight degenerate because man has two eyes and a relic instead of three? Certainly not! Then why should his sense of smell be termed degenerate, merely because man now has two olfactory clefts and the remains of an organ of Jacobson?

The one distance receptor in which there are definite signs of degeneracy is the ear. Inside its rim is Darwin's point—vestige of a pointed ear. Also, man has the remains of ear muscles, as apes have.

But these things—pineal gland, organ of Jacobson, ear muscles—matter little. The true test of a sense is in the organs in use, not in the useless relics that remain.

THE NUMBER OF PAIRS OF CONCHAE PRESENT: First of all, we note that these lateral ridges per se have nothing to do with the sense of smell, for they are present all through the nasal cavity, even where there is no olfactory sensitiveness.

Also: there is no direct parallel to be drawn between the number of pairs of them present and the olfactory acuity of the different species of mammals. That is, an animal with ten pairs has not twice the acuity of one which possesses only five. As Parker himself states, their number is used simply as an identification of a division already made—presumably by common sense.

My dog roots in the ground for a buried bone. His nose, softer than the dirt, is pushed into wrinkles, lateral folds like the conchae he already possesses.

And there's our clue! The ant-eater roots in the ground continually. It has ten pairs of conchae. The hoofed mammal, which gets its food by cropping low-growing herbs and grasses, has eight. Checking through the feeding habits of the various species, we discover that the more earth an animal comes in contact with in finding its food, the more wrinkles—conchae—it has in its nose.

There is a striking resemblance between them and the dust-boxes to be found in any airshaft, and this suggests their true purpose. An animal which finds its food on or under the ground where dust or sharp grass blades may enter the nostrils and irritate tender membrane—needs more of these ridges to protect itself.

If we state that possession of more than four pairs proves that the animal gets its food on or near the ground, and possession of fewer proves that the animal gets its food elsewhere, we have said all that seems justified about these ridges. Grazing animals, dogs, ant-eaters have more than four pairs. Men, monkeys, seals and whales have less.

But directly or indirectly, these conchae offer no evidence as to the keeness or degeneracy of man's or any other animal's sense of smell.

THE COMPARATIVE OLFACTORY AREAS: Whatever proof science has to back up its dogma of degeneracy must be found in the fourth statement: that comparatively the olfactory area in man's nose and brain is smaller than in the keener-scented animals.

So too has man a smaller eye, relatively, than many of these same mammals, yet it has been proved that man has the keener sight. Consider the Pekingese dog, for example. That lap-hound has eyes much larger, comparatively, than man, but it lives in a "dim, colorless world of sounds and smells." The humming-bird possesses a brain three times as large, in proportion to its body weight, as the human brain—but a humming-bird has never heard of inductive logic. Take the size of a man's brain with reference to his ability to think. Turgeniev, talanted enough, but scarcely the world's greatest genius, had the third largest brain ever measured—but one of the two larger than his belonged to an idiot. Finally, a man's brain is larger than a woman's, but does that prove him more intelligent? Eight hundred million women will back me—and the scientists—in saying that it does not.

The size of an organ has nothing to do with its importance—or its functioning ability. The comparative size makes little difference, either.

The shark has taste-buds all over its body—and yet a shark can scarcely be considered a gourmet.

To put it playfully: a big-footed man is no better walker than Cinderella. He simply covers more ground—as he walks.

Condition is much more important than size, for an organ need only be large enough to do the work it has to do, and the rule is absolute that every body organ, no matter how small, has this ability. Almost, it seems that the smaller the organ, the more important its task. Cut off a man's legs at the hips, and he'll live to sue you. Cut out one of the tiniest glands in his body, and he'll die in two hours.

Condition is what counts. How healthy is the organ? how well-fed? how well-served by nerves? how well-equipped to do its work?

To every question the sense of smell gets an O.K. from physiology. Much more rarely than any of the other special senses does it lose its health. It is better fed than any other organ in the body. Nerve connections? It is *all* nerves. And it is so well-equipped to do work that every authority expresses wonder that it apparently does so little.

Some writers, after describing newer, more intricate receptors, speak almost contemptuously of this old, primitive organ. It is true that the smell-sense is, after touch, the most ancient of the senses. It is equally true that there is an almost primeval simplicity about it as compared to the so-called higher senses. But what do age or simple structure have to do with degeneracy?

Nothing!

Which leaves us—where? Back at the beginning again, and the common knowledge that man has lost his sense of smell.

We have determined nothing, positively. If man does not use his sense of smell, he might as well not have it. And yet—here is the sense, to all intents and purposes intact, ready for use. It seems amazing that man—a puzzled, frightened child in a tremendous, unknown world—man, who needs every possible means of communication, should ignore this sense, should hobble along on three senses instead of four.

Amazing indeed, which is why, perhaps, Dr. Ponder says: "Smell is a little island untouched by science. We don't know how it functions. We don't know how to use it."

In sum and substance, science knows what has been set down so far. It is such a very little, isn't it! . . . but we'll call it our first chapter.

Science has not troubled to doubt the statement that man has lost his sense of smell. No one, apparently, ever has doubted the statement. There are, however, no physical signs of degeneracy in the organ itself. So the matter rests, finally, on common knowledge and observation of how little man, in comparison with known keen-scented animals, uses this sense.

Together, I think, we shall jam a new science into this common knowledge.

CHAPTER TWO

AT THE BEGINNING, there's a small difficulty. We must come uncomfortably close to a taboo. We are still on firm ground; what follow are irrefutable facts; every step ought to be smooth travelling.

But still—that taboo! Why won't your best friend tell you? Why are girls so unwilling to be bridesmaids these days? . . . Maybe we'd better talk about animal behavior a little, first.

A certain species of ant, Fielde says, bears three distinct odors: a nest odor, a family odor, and an individual odor, deposited by her own feet, whereby she traces her steps. Lubbock returned ants to their nest after a separation of a year and nine months—yet they were amicably recognized as friends. Ants—and in this they have a priceless possession man lacks—recognize their friends even when intoxicated, and know the young born out of their own nests, even when the young have been taken out of the cocoon by strangers. Wheeler says that even the degenerate human nose can detect the different species and castes of ants.

. . . All of which redundantly demonstrates the admitted fact that ants have an odor and a sense of smell.

McIndoo, working with bees, recognized a number of characteristic bee odors: the hive odor, the brood odor, the honey odor, the pollen or bee-bread odor, the wax odor, and the odor coming from the bee's sting. He found that the drones emit an odor peculiar to their sex, and that each worker gives off an individual odor different from that of any other bee. When a bee finds a bush or bed of flowers in full bloom, it returns to the hive impregnated with the smell. The other worker bees seek the same flower-scent, disregarding all others, in order to reach the same treasure-trove.

So, we find, bees, like ants, have and are affected by smell. And throughout the insect kingdom, the same two statements are true.

Despite the fact that birds possess a well-developed olfactory organ, doubt has been expressed that they have an acute sense of smell. Darwin proved to his own satisfaction that Andean condors, at least, have poor olfaction. And yet—crows and vultures scent a battlefield from afar. Buzzards and carrion crows, in the South, will assemble from miles around at a closed barn in which a cow lies dead. Rooks and woodpeckers can detect grubs beneath the bark of trees. Petrels can smell offal thrown from boats in the mist. Other sea-scavengers have the same faculty. In the Norfolk game decoys, the watching decoyman commonly burns peat to prevent the geese and ducks from scenting him downwind. The great bustard deserts the nest if the eggs have been handled. So do the pea-hen, the guinea fowl, and the quail.

Certainly birds have an odor. It seems almost certain that they have a sense of smell.

The toothed whales and porpoises, it is said, have a very poor smell-sense. I have been unable to find an account of the experiments—if any have been made—which prove this. And it would seem at least diffi-cult to complete any truly exact and scientific research on such animals. Imagine a Dr. Laird engaging in his class-room experiments—with a school of whales!

It is more likely that the assertion is based on the cerebral anatomy of the creatures named, as is done in the case of man and his ape-cousins. If so, any doubt thrown on man's smell-lessness should apply with equal force to these animals.

The whales, porpoises and seals have however a reason that man does not possess for being anosmic. For they, in common with all the living world of earth and air, left the sea millions of years ago and, as they de-veloped lungs with which to breathe, so did they modify their organs of smell to receive sensations from air, not water. Eons later, in compara-tively recent geologic times, when these few mammals left the land and returned to the water, this differentiation for air-smelling had gone too far. Their senses of smell failed to react to water-borne stimuli. So we see, in their living descendants, the organs simply atrophied—even to complete obliteration in the dolphin.

It is worth while discussing these anosmic mammals, and what caused them to lose the sense of smell. For poor olfactory powers are attributed to so few creatures. Man, other primates, whales, seals, porpoises, and some birds—the smell-lessness of the last being highly controversial—here is the complete list. Even those creatures that dwell in the swirling currents of a medium that itself must possess a most penetrating odor—water—even those admittedly have a sense of smell—and use it.

You have been told that pure water is odorless. It is not. The western burro, left unhobbled, will never die of thirst if there is any water within miles. A thirsty traveller, lost on the desert, is always advised to let his horse lead him to the nearest water-hole. Hudson, in his "Naturalist in La Plata," says that horses can smell water from a distance of thirty miles—and cattle can detect its presence even farther.

Yet in water, among certain fishes—the catfishes and dogfishes, for example—feeding scarcely ever occurs, even though the fish are starv-ing and food is present, unless the fish smell the food.

The sea-anemone uses its sense of smell to find food and to avoid danger. The barnacle uses the same organ to find the ship it attaches itself to—and it has been said that if a marine paint could be discovered, the odor of which was offensive to barnacles, ships would never need go into dry dock, and barnacles would lose the distinction an ocean voyage gives them—and their human equivalents.

Crabs and lobsters use smell to an extraordinary degree. Ehrenstein notes remarkable examples of the achievements of the sense in amphibians and reptiles. And finally, as we've been told, the sense of smell finds its highest development in mammals.

In *all* mammals, including some that you would think use their noses only to sneeze. Darwin tells of a blind coach-horse which stopped at every inn, whether the coachman wanted to or not.

The sense of smell is a universal sense, for with the single exception of the mammals which have gone back to the sea, we find it alive and in working order in every living creature. And, with remarkably few exceptions, it not only is in working order—it works!

Why is it so universally present, so (almost) universally used? The answer is plain: there is such need of it. Everything has an odor. For odors are chemical particles, and whatever else life is, we know that it is a chemical laboratory. And wherever we see life, we know that there is ceaseless chemical action with, ceaselessly, particles therefrom diffused into the surrounding medium. A constant by-product of living is the odor caused by the chemical action which sustains life.

Theophrastus said that every plant, animal, or inanimate thing has a scent—and one peculiar to itself. Passy states that he has never met with any object that is really inodorous when one pays attention to it—not even glass.

There is no need, here, to go so far. But it is very certain that every living creature, every animal has an odor.

And so, after beating around a sequoia-like bush, we go from the logical general to the particular—and say that man has an odor.

A race-odor. The smell of the Negro to the white, it is universally known, is unpleasant. Not so well known, perhaps, is the fact that a white man's odor—to the Chinese, Japanese, Hindus, and certain Negro tribes—in other words: to considerably more than half the world, is equally unpleasant. The smell of Europeans, as Adachi, Japanese anthropologist, describes it, is a "strong and pungent smell, sometimes sweet, sometimes bitter, of varying strength in different individuals." The Negroes perceive in the white man the smell of putrefaction. The Hindus, not polite either, describe it as a smell like that of rotten meat, and say it is caused by the whites being meat-eaters.

In contrast to this is the curiously ethereal smell attributed to the

Hindus by certain white persons, who credit it to the Hindus' meatless diet. Strange company! The Chinese, Negresses and blondes—all are supposed to have the smell of musk. Red-haired women, judged by odor alone, are a race apart . . . And so on.

One acute thinker declares that racial antipathy, which causes so many wars, draws its source from differences in race-smell.

And this odor is not the result of failure to wash. It is *increased* by cleanliness. Havelock Ellis states that this is true of Negroes. Adachi asserts that it is equally true of whites.

In addition to the smell of the race to which he belongs, man has another specific odor, one peculiar to himself. As one writer puts it: "Each individual has his or her own odor as distinct and personal as is his countenance."

There is ample scientific reason for this. The products of metabolism which we constantly exude through the skin are odorous. Since no two of us are chemically alike, in no two individuals are the products equally alike and in exactly the same proportion. And not only the surface of the skin, but every pore contains these odorous particles.

We call these pores the sweat glands. There are, in addition, sebaceous glands, the primary function of which is to lubricate the skin, but which also have a secondary use as a means of recognition. The existence of a third type of special scent-making glands has been suspected, though never proved.

It must be remembered that the general skin odor and the odor of the arm pits are after all only two of the components of an individual's scent. Havelock Ellis divides this composite aura into nine components. They can be grouped more euphemiously here into six: skin odor, arm-pit odor, hair odor, mouth odor, the odor of the breath, and emanations from other bodily openings.

The combination of these makes each individual's distinctly individual odor. *Your* distinctly individual odor, reader! And whatever slight shock you feel is the taboo's outraged protest.

In passing, it may be remarked how a certain corporation profits hugely by whispered screams about this taboo. Remember in frequent full page ads how terrifying initials have a following asterisk that guides you to a tiny explanation at the bottom of the page, where you learn that B.O.* means—whisper it—body odor? The corporation refers in these same ads to but one of the six components of your individual

---

* A plus B, in algebra and in life, equals A plus B. And "B.O." plus a smelly soap equals body odor plus the smell of the soap—but we're wasting time.

fragrance and claims to destroy that one—whereas, actually, it heightens it.

And the halitosis that makes you often a bridesmaid, etc., may possibly, if you are a mouth-breather, comprise some five percent of the sweet fragrancy that is *you!*

Any dog can tell you so. And any dog knows his master or mistress, washed or unwashed. And *my* dog seems to think a certain widely advertised soap has a smell of its own which my dog, who can't read advertisements, poor fellow!—considers almost as unpleasant as the smell of another just-as-widely-advertised mouth wash.

If you have a smell to an animal—you have an odor. In tests in Germany, after the war, a police dog was able to follow his master's scent even under snow. Instances are on record of a dog which was able to find out of a heap the one stone that had been thrown by his master. Bits of wood were held in forceps. The dog was able to detect the piece that had been touched by his master, even when the wood had been touched by the finger-tips for only a second or two. He succeeded in identifying the piece even when persons other than the owner who was to be identified had handled the wood too, and even after artificial odors had been daubed on it, although in this case he barked his disapproval. It is difficult to blame him ... Dogs have a most curious antipathy to synthetic scents, anyway.

There are records of men and women whose natural fragrance was so penetrating that all around them were conscious of it. One such was Alexander the Great, whose tunics were drenched with the natural perfume of his body, who, as Plutarch says, was loved by women more than any other prince, because he smelled so sweet. Malherbe, Cuvas, and Heller had a musky odor. The good gray poet, Whitman, diffused an agreeable scent.

An odor of sanctity is stressed in all the legends of the saints. The modern explanation offered for this opens the gate for our next observation: that a man's personal odor is not a fixed and unvarying thing.

Havelock Ellis writes: "The odor of sanctity was due to abnormal nervous conditions, for it is well known that such conditions affect the odor, and, in insanity for instance, the presence is noted of bodily odors which have sometimes been considered of diagnostic importance."

Thompson says that the odor of sanctity in which the saints died may have been simply the smell of the mortal disease from which they suffered, and that the sweet smell credited to Saint Theresa was almost certainly caused by her malady, diabetic acetonaemia.

The characteristic smell of various diseases has long been recognized. Persons with a suppurative condition give off the odor of strawberries, caused by the bacillus pyocyaneus in the pus. Persons to whom turpentine has been administered emit the odor of violets. Smallpox has a characteristic odor. Pellagra has the smell of sour bread. Sepsis, diphtheria, and measles have each their peculiar scent. Chronic cystitis emits an ammoniacal odor ... And so on.

The fact that odor changes in sickness is interesting because we are, none of us, well. We are walking hospitals. Another analogy would be to compare our bodies with Germany in the late war—besieged on all sides, losing men daily, trying to raise the birth rate to provide more troops to face the guns. The white corpuscles in our blood become so worn out by the continuous warfare that they die in ten to twelve days, and our bodies must continually manufacture new ones to act as shock troops against the deadly little enemies we inhale or swallow or let slip in through an unguarded break in the skin.

In direct proportion as these engagements are pitched battles or skirmishes, so do they affect certain components of the scent that is as peculiar to each man as his finger prints. Any serious derangement of the body will of course cause a tremendous change in this odor. Then we have the characteristic disease smell. But even the minor illnesses have their appreciable effect. Perhaps it will be simply a slight difference in mouth odor—an increase in the carbon dioxide content of the exhaled breath, perhaps merely a heightened activity of the sweat glands —but there must be a change. For change in the chemistry of any part of the body must mean a change in the chemical particles thrown off by that body.

Since most of the mouth odor comes from the stomach, the food we have eaten causes such change. The scent of onions may be detected in the blood four days after the eating.

We might even say that the odor of food not yet consumed may make a difference. For certain glands begin to discharge their secretions the moment the nose smells food. The acts of chewing, of swallowing, of digesting all cause the release of chemicals into the alimentary canal, the lymph system, and the blood stream.

Fatigue (because of its waste-products), even the kind of work done may likewise change the personal odor. Alfred R. Wallace wrote of a dog's antipathy to butchers, and suggested that some peculiar aura is developed in human beings by constant contact with flesh.

Sickness, mental or physical; ingested food; fatigue; environment— to these must be added one other odor-affector: emotion. I remember once sitting on a couch after a harrowing domestic battle which I had, naturally, lost. I was trying to read a book, but I was so full of bottled-up anger that the page seemed to be printed in red. My dog, who had been absent on personal business of her own, came jog-trotting into the room. She started over to tell me about it; came half-way; thought better of it, and stopped. With her head on her paws she lay watching me eagerly, but very doubtfully. And she whined.

Now my dog, who is a collie, believes that I, though a painfully dumb scholar, still have acquired at least the rudiments of dog language, and it is her habit to come to me whenever she returns from one of her ventures into the wide world and tell me all about it.

Something about me, this time, had stopped her. I had not spoken. I am very certain that I had not moved, and I have no bright and shining countenance at any time. And yet, something had warned her not to disturb me. What had it been?

Every dog-owner will concur in this answer: a dog knows his master's moods.

In this case, how? By sight? To the dog who, as Herrick tells us, has no true color vision, no real power to discriminate fine details of form? By hearing? There had been no appreciable sound.

The answer is: by smell, of course. And in this example of a dog's nose, everyone is willing to admit it. There is even a scientific explanation offered: emotion caused a ductless gland to release adrenalin into my blood. The adrenalin in turn caused certain other visceral changes—such as releasing sugar into the blood from the stored glycogen in my liver. And the odor from these chemical actions necessarily changed my personal odor to an extent.

To quote: " We constantly exude products of metabolism through the skin. The nature of these secretions is modified by different strong feelings and emotions."

To that dog of mine "living in a dim gray world alive with odors," there was a perceptible change in the most important of all her odor-sources—me! To her nose, I was still recognizably her master, but I was the familiar smell *plus* something—a strange, terrifying smell of anger. And so she lay on her paws—and whined.

Certainly the truest proof of an odor or any change in such odor will be found in animal behavior, specifically: among the macrosmatic animals, whose sense of smell is undoubted. A kennelman will tell you that a dog never bites a man who has no fear of him. A bee-keeper will make a similar statement about bees. And very firmly I believe both experts. It happens that I have no fear of dogs. I can stop dog fights;

I can meet wolf-souled police dogs, unstirred, unafraid, and unbitten. But I have a real fear of bees—and I am stung by them, frequently.

As a boy in the South, I owned a hive of the things. A playmate, a little Negro boy, used to rob it for me every spring and fall. Wrapped in mosquito netting, waving two rag torches around my head, and always a hundred feet away at least, I used to superintend the job. The black boy's only protection was his absolute fearlessness of bees. Invariably, he would go unscathed; I would be stung a dozen times. It was embarrassing to a young Nordic—or would have been, had I not also possessed a pack of seventeen halfgrown hound puppies, and the knowledge that my little playmate had the marks of old bites all over him—and a mortal fear of dogs. And so, invariably, after the honey was safely gathered and my little friend had started up the road, too, too loudly happy over the bee-stings I had received and he had not, I released one or two or three of the dogs, as many as were necessary— and youthful white supremacy was again saved.

I knew an old Negro hunter who had a trained raccoon. The little animal lived among the coon-hounds of the pack, strictly on sufferance, and accompanied them and the Negro on hunts, trailing with the dogs until it got tired, then riding on the old Negro's shoulder or saddle-pommel. The game hunted was always other coons, and whenever one was treed, the little cannibal coon would slip off the saddle, through the howling pack of dogs, and into the hollow of the tree, eager to do battle with its wild brother.

(By the way, this is a true story.)

Inside hollow log or stump or tree, the war was fought, and always at the end the two coons came out. Usually the tame one was the victor and came out last, but anyway, out they always came. The huntsman held his fire to make sure he wasn't shooting his pet; the dogs always chose the right coon to chase, and the old Negro, because he had no trees to cut down, was killing more varmints than any three hunters in the woods.

Disaster came. A strange dog, hunting on his own, joined the pack one night. The two coons came out. The strange dog chased the wrong one. The pet coon scurried away for its life. The old Negro blasphemously drove the strange dog away and turned around—to find that his own pack had killed the pet.

To say that the dogs killed the coon because it ran would not cover the case, for the pet coon had run frequently—before or after its wild relation. The only reasonable explanation is one of fear. The pet coon, frightened by the menace of the strange dog, bore the peculiarly changed coon smell the Negro's hounds had learned to associate with wild coons—their permitted prey.

Even without confirmation in animal behavior, there seems no questioning the fact that every human, laboring under stress of emotion, undergoes visceral, chemical changes. And no matter how small these are, there must be corresponding change in the body odor that is the resultant at any moment of all the chemical processes going on within that body. It is a matter of simple arithmetic. These changes *must* occur.

My dog knows me from all the world whether I am clean or dirty, hungry or fed, glad or sad, dyspeptic or at peace with my innards. Admittedly, the changes in the individual's odor are normally small. But they are changes, and they do make a difference, though usually not too tremendous a difference. As an analogy, consider the possible alterations in a man's personal appearance: shaved, unshaven, dirty, clean, dressed, or in the buff. He's not always the same—but he's still recognizably himself.

To sum up: every human has an individual aura, peculiarly his own. This aura, which is the resultant of certain lesser odors caused by the body's functioning, and arising both inside and on the surface of the body, may be changed to a degree by the physical condition of the body, by food in the alimentary canal, and by emotional disturbance.

Finally: remember that this personal, variable aura is all around and about its author; is in the air around him; part of the air he breathes. Certain components of it: the odors from the alimentary canal and the lungs, are even part of the air he breathes out. It is with him always—walking, standing, sleeping—*every moment of his life.*

CHAPTER THREE

G. K. CHESTERTON made his dog Quoodle sing a song about the nose-lessness of man. But, says H. Stanley Redgrove, in SCENT AND ALL ABOUT IT—somewhat too inclusive a title—the chemist and the perfumer are certainly not noseless. For their highly trained noses enable them to differentiate between a multitude of odors, identifying substances by this means and even detecting the different constituents combined in a smell. Remember the story of Frangipani, the Italian perfumer who accompanied Columbus on his voyage. He it was who told the despairing others that the perfume which was wafted to them, approaching the unseen isle of Antigua, came from flowers—sure proof that land was near.

The making of synthetic perfume is a great industry, and an exact one. Quantities of various substances are measured with the utmost precision; temperature and time are watched ceaselessly. And yet, at the end, after every other form of test says that the mixture is right, it is submitted to the most important man in the establishment, and he tests it—with his nose. If *it* tells him that the stuff is wrong, the whole batch is discarded—or worked over.

His sense of smell is more exact than any man-made testing instrument. Even the title given him in the factory recognizes this. He is the "nose."

In the great natural perfumeries, there are other such "noses" to keep the blends the same. Of these men, there are two or three or four great "noses" who can recognize—the figures are exact—two hundred perfume odors.

Two hundred! The ultimate in perfumes! The stock example of a trained sense of smell! . . . It's worth remembering, that number.

The little Armenian who rings your doorbell can tell—by his nose—which rugs in a shipment are developing dry-rot—and sells these first, of course. The textile expert can determine from the odor of burning cloth whether it is a vegetable or animal product. The pharmacist uses his nose in compounding a prescription; the bacteriologist uses it in isolating his cultures. The North Atlantic seamen (although this may also be caused by sensation in the cold spots of the tacile sense) are able to smell the proximity of icebergs. The story persists that the sailor on watch aboard the Titanic reported smelling ice to his officer, shortly before the crash, but the warning was disregarded because the ship was farther south than icebergs had ever before been known.

McKenzie found that the Indians could smell the presence of a camp-fire farther away than he could see it. A trading schooner dropped anchor off the desert coast of Australia. Water had run out three days before and the men were dying of thirst. No one on board had ever seen the spot before. And yet, according to the British captain, his Solomon Islander crew led him straight across country, up hill and down, four miles to a tiny water hole. They told him they *smelled* the water.

Authentic cases prove that certain persons can smell the approach of rain. A writer quoted in the Literary Digest, admitting the fact, offers an involved explanation regarding the release of noxious gases in near-by swamps caused by the decrease in atmospheric pressure. The gas, he says, is what these hypersensitive folk smell. I have met one or two such persons and it happens that where they lived, there were no swamps. A much more reasonable explanation is that it generally rains somewhere else first and the rain approaches with the wind, but more slowly. The rain falling elsewhere is what the rain-smeller senses. And there is no great need for any hypersensitivity, either. Anyone can smell rain or, at any rate, sense a new, fresh smell when the rain is near enough to be seen approaching.

Dr. Curran Pope says that doctors should again avail themselves of the sense of smell, once widely used in diagnosis. Dr. Mayo speaks of its value to a surgeon.

"My father," a medical friend writes, "was a doctor of the old school, and he used his nose. Several times, while walking down an unfamiliar street in a strange town, he has pointed out a house to me and said: 'There is diphtheria in that house.' And he was always right."

The judges of jelly, etc., who make one housewife happy and another miserable by the award of prizes at county fairs do not eat the jelly. They *smell* it! And on what their noses tell them, they base their decisions.

In every metropolitan hotel there is a chef whose sole duty is to make salad dressings. Too much or too little of any ingredient would ruin the sauce, and lose the chef his job, and so he has to be sure it's right. Before you begin pitying a poor devil who must spend eight hours a day tasting salad-dressings to earn the daily bread which so much mayonnaise makes him too nauseated to eat, listen: he never tastes the dressing at all. He smells it—and his nose tells him if it is right or, if not, what to add to make it so.

And many another person has been known to smell sugar in tea and salt in soup!

Taste, you remember, except for the four gross distinctions of the taste-buds, is wholly smell. There are professional food-tasters by the hundred employed in dairies, in soup factories, in fruit and vegetable canning establishments.

Remember Porthos, one of Dumas' "Three Musketeers," who could tell the variety and age of the wine he drank. Every connoisseur prides himself on this ability. Watch a Frenchman with his small-necked mug of Napoleon brandy—warming it with his hands, savoring the bouquet of it for hours, almost—and you'll come to believe, as he does, that the actual drinking is an anticlimax.

There's an old story of two Spanish wine testers and a just-opened hogshead of port. Both smelled the wine, tasted it, and both agreed that it was bad. One said it had a metallic taste; the other averred that the smell of leather spoiled it. Both were experts, and the dispute grew strong. Finally the wine was poured out, and in the bottom of the cask was found—a small leather-covered nail.

The story is so near the truth. . . .

Half the pleasure of life is in eating, and much of the other half comprises different ways of developing an appetite. Thank your nose for this! Your nose selects the viands and savors them. Your nose tells you whether you ought to fire that cook or marry her.

The tea-taster, after sampling and blending teas from half the plantations of the world, giving you the pedigree of each leaf, its country of origin, the season of the year it was picked, at what altitude it was grown, and even perhaps the very garden it came from, will take time off to tell you that you can't get a good cup of tea in America, that the sea air—no matter how perfectly the tea is packed—always spoils the flavor. And to him at least, with the unerring tea-consciousness of his nose, it does.

You've been told that if you hold your nose closed, you cannot distinguish by the taste buds alone between black coffee and quinine. Well, the professional coffee taster, leaving his nose open, can discriminate between seven hundred varieties and blends of the brew. Seven hundred! . . . Another figure for us to remember.

The sense of taste, in its nuances, is the sense of smell. The olfactory region of the nose is like a room with two doors: an inner and an outer one. The dog has the same two doors—and yet my dog has no confidence whatever in *her* sense of taste—and a great deal in mine. Whatever she sees me eating, she wants. If she reasons at all, she reasons that no matter how curious it tastes, if I'm eating it, it's good.

No one has ever said that a dog's sense of taste is better than a man's. And no one has ever said that primitive man had a better taste-sense than we have.

A room with two doors!

If the sense of smell has not degenerated as regards its power to taste, if the man of today gets every bit as much pleasure out of the flavor of food as his tree-dwelling ancestor obtained, if sensations via the rear entrance to the nose are as strong as they ever were—then this question of degeneracy may be summed up in an apt analogy:

You go to a physician for treatment. Outside, you notice that there are two doors to his office. Someone tells you that if you enter by the back door, you'll find the doctor a skilled physician; if, however, you enter through the front door, you'll find him no good at all.

There's something absurd about it, isn't there?

We'll go on knocking at that front door.

William Beebe says he finds that something always makes him leap aside in time to avoid being bitten by poisonous snakes in the jungle. Reluctantly, apologetically (that's the taboo working, even on a scientist) he tells us why. He believes he smells them—unconsciously.

Lumholtz, while living with the cannibal and ophiophagous (snake-eating) tribes of Queensland, found that they hunted the snake, a large species of boa on which they fed, *by its scent.* The serpent travels long distances in quest of prey, and the natives, once on its scent, would follow it like a pack of beagles, through woods and thickets and marshes, over rocky tracts and all kinds of country until they caught up with it. The scent, they assured the naturalist, was strong and easy to follow.

Like a pack of beagles!

The Peruvian Indians follow game as a dog does, by scent. The Solomon Islanders sense, through the night and miles away, the presence of an island "to leeward," and know what island it is. They say they do this by smell. The shamans of the Siberian wastelands track bears long distances by this same sense.

Certain civilized persons have a peculiar antipathy toward cats. They are able to detect the presence of one in a strange room, and even in the dark point out where the animal is. Unless one wishes to attribute this cat-phobia to some form of occultism, the only possible conclusion is that these persons are able to smell the beasts.

Other instances of such curious aversions to particular smells are recorded. Madame de Staël fainted whenever a lobster was brought to the table. The odor of horse-radish has driven men to the verge of madness.

Hudson tells of a sick woman, bedridden with bronchitis, whom he visited. In the room were two stinking dogs, to which the woman paid no attention. A maid entered—and remained in the room a minute and a half. The sick woman immediately insisted that she could not stand the girl's smell.

Red-headed women are notably odorous. A young man in Paris felt a strange attraction towards them, and, instead of concealing this attraction as most men do (the married ones, anyway), he gave free rein to his nose. He was able, through his sense of smell, to detect the presence of a red-haired woman unseen to him.

Admittedly the last few examples prove nothing, except perhaps that the persons described were abnormal. But instances of normal men recognizing their fellows by odor alone are indeed important.

Remember, in the second chapter, the various examples given of race smell, white, black, and yellow? How were such differences determined? How was Adachi able to describe the odor of the European white? Or Ellis, of the African black? Because: men were able to sense these differences.

The Nicobarese, according to Man, can distinguish a member of each of the six tribes of the archipelago by smell. Emin told Mungo Parke that he could pick out members of different tribes by their characteristic odor. The Ethiopians can distinguish friend from enemy in the dark, by smell. The "History of the Antilles" assures us that some of the Negroes there can, by scent alone, distinguish the footprints of a Frenchman from those of one of their own race. The remarkably keen sense of smell possessed by the North American Indians, so McAdoo says, accounts in part for their wonderful power of trailing enemies. Humboldt tells us the Peruvian Indians can distinguish between the footprints of their own people and those of whites and Negroes. The Australian Bushmen are said to follow the scent of man, as a dog does. Another writer says he repeatedly conducted experiments which proved that Negroes and Indians recognize persons in darkness by their scent.

Dr. C. S. Myers of Sarawak noted that his Malay boy sorted the clean linen according to the skin odor of the wearer. Chinese servants are said to do the same, and so do Australians and natives of Luzon.

You will note that this concerns the personal odor of individuals, and the fact that the linen was *clean* jibes neatly with the assertion already made that body odor is increased by cleanliness.

Governesses have been known when blindfolded to recognize owner-ship of their pupils' garments by smell alone. Havelock Ellis remarks: "Such a case is known to me." There are some Europeans, he adds, who can recognize and distinguish their friends entirely by smell. The case has been recorded of a man who, with bandaged eyes, could recog-nize his acquaintances at a distance of several paces, the moment they entered the room.

Years ago, in geography class, I was highly amused to learn that certain savage races rubbed noses in greeting. What savages they were! Everybody knew that the sensible way was to shake hands. At the time, I was not aware that the civilized gesture arose from the an-cient custom of holding out the hand, empty, to prove to your equally civilized neighbor that you weren't concealing a knife—and murder-ous intentions.

According to Lewis: in one hill-tribe in India, they sniff at a friend's cheek and say, by way of salutation: "Smell me!" On the Gambia, according to F. Moore, a man saluting a woman, instead of shaking hands, puts the woman's hands to his nose and smells the back of it twice.

The smell greeting, actually, is employed over nearly half the earth's surface. The nose is applied either to nose, face, or hand of a friend in salute. The usage is common throughout a large part of the Pacific and elsewhere, among the Papuans, the Eskimos, the hill-tribes of India, and others.

Savages all, of course.

St. Phillip Neri is said to have been able to recognize a chaste man by his smell. There was a monk at Prague, who unfortunately died just when he was about to compose a new science of odors. He could recognize by smell the chastity of the women who approached him.

What wouldn't they pay him in certain Fraternity Houses today!

Nearly all the poets and novelists have been strongly affected by and interested in odors, including human ones. Shakespeare, Goethe, Baudelaire, Herrick, Dickens, Huysmans, Tolstoi, Schiller, Balzac, de Maupassant, Stevenson, Meredith, d'Annunzio—the list is almost endless.

And thus we have the rather amazing paradox of men at either end of civilization—snake-eating savage and genius-poet—using a smell-sense that civilized man almost angrily denies possessing!

Although he is perfectly willing to grant that the savages, at least, possess it. McIndoo states: "The smelling-power of these primitives is inconceivable to the average civilized man. . . . Among many savage tribes, the sense is as acute as in many of the lower animals."

When we consider that by this sense the savages are able to find food, hunt game, recognize friends and enemies, do everything an animal's nose can do—we in honesty admit the statement.

By simple definition, then, these savages are macrosmatic. And what about the number of conchae, those ridges which are the criteria, so-called, of olfaction? And what about the size of the olfactory area?

This: *The savage tested anatomically, has exactly the same organ of smell as other men.* He possesses, like his white brother, four pairs of conchae or less. And his olfactory area—like civilized man's—measures approximately one square inch.

And perhaps this will prepare you for a rather extraordinary fact: *The savage, tested objectively, has no better sense of smell than a white man.*

What about the hypersensitives of the olfactive type: the sufferers from cat-phobia; the young man who smelled red-haired women, and so on? This: *The hypersensitive, tested objectively, has no better sense of smell than an ordinary man.*

Tradition lingers of the marvelous olfactive powers possessed by the Japanese. *The Japanese, tested objectively, has no better sense of smell than the white man.*

There is a universal belief that Nature compensates the unfortunates who have lost one or more of their senses by making the remaining ones more acute. You've read about the deaf and blind Helen Keller, and her marvellous sense of touch, her wonderful sensitiveness to odors. *Dr. Tilney found that when tested objectively, Helen Keller's olfactory sense shows nothing above the normal average. The fundamental pathway for the sense of smell in Helen Keller has absolutely no advantage over that of the normal individual.*

And that's that! If by this time you are in a state of bewilderment as to what the whole thing means, you have attained the exact present scientific attitude towards the sense of smell. Science avoids it like the plague. The sense seems to be lawless as a moss-trooper, mad as a hatter.

By the very definition of odorous substances, any volatile gas ought to be odorous, and yet to man's consciousness, many of them are not. All odorous substances, exceptions excepted, belong to the fifth, sixth, and seventh groups of Mendelief's system—and so do a great many other substances that have no odor. Man, we are told, has lost most of his sense of smell—and yet man can smell impurities in the air in such infinitesimal proportions that no man-made test can detect them. If odor is a chemical sense, things chemically alike ought to smell alike— but they don't. If odor is a wave-sense, things with the same wave-length ought to smell alike. Again, they don't!

Ponder termed smell: "a little island untouched by science." That's not true! Science has touched it; landed on it; claimed it by right of discovery and colonization scores of times. Two thousand years ago Galen, who never knew vivisection was wrong, performed a Caesarian operation on a goat heavy with young and removed a living kid. He put the kid down before a row of pans, one full of oil, another of vinegar, a third of milk, and so on. The kid moved along the row smelling each pan—until it reached the milk. And there it stopped, and made its first meal! Galen remarked that this proved the existence of instinct, and since the kid was removed before the natural time of gestation, and had never even seen the outside of its mother, it sounds reasonable enough.

The point is that here, at the very dawn of science, one of the earliest scientists was observing the sense of smell and drawing conclusions therefrom. The analogy: "a little island untouched by science" sounds a little too much as if the question would be settled in a moment, once Science found time from more important things to attend to so trivial a matter.

... Which is—sour grapes!

Let's make a better analogy, a truer one. Smell is an uncharted shoal, on which every bark sent out by science has so far been wrecked.

Honest science admits it. And when science admits failure, personal knowledge and belief need feel no shame in giving up, also. And so I ask you, for a little, to forget what civilized man can and can not smell— what *you* can and can not smell—and look at the matter objectively.

If Henry Ford builds a vehicle, tries it out on proving-ground and open road and discovers: that it can be guided by a steering wheel in the hands of a driver; that it has a speed of sixty miles an hour and that it can take oil and gas, and of them make miles traversed, he is justified in assuming that a like vehicle, measuring the same throughout, given the same fuel, and an equally good driver, will render exactly the same performance.

If one human is able to hunt food, to follow a scent, to recognize friend from enemy, to distinguish one person from another entirely by the sense of smell—should not another, with exactly the same physiological equipment, have the same ability?

Possibly the objection arises that I am comparing men as though they were machines, and that the individual human can not be so compared because he has certain private intangibles that a machine has not: consciousness, memory, reasoning power, neuroses, romances, and so forth.

It takes a complete human to have these things. Man's various parts, organs, senses are simply machines without these powers: the eye being simply a camera; the heart, a pump; the lungs, a bellows; the olfactory area of the nose, a chemical reactor.

However, let's leave the individual and take the tribe, or race. If Peruvian Indians have a sense of smell approximately that of the macrosmatic animals, should not civilized Americans, with exactly the same physiological equipment have exactly the same power?

The answer once more, for emphasis, is that tested objectively, civilized New Yorker has as good a sense of smell as the Queensland savage hunting snake for his Sunday dinner.

If I add: "—but civilized man does not use it," I can quote authority, book and page, for the complete statement. I do not think though that this is the whole truth.

Two vehicles with the same equipment should give—and will give—the same performance unless, in the case of the second vehicle, its driver is forbidden to drive, or there is a writ of prohibition against it, which keeps it from moving.

A writ of prohibition! We'll get to that in a little.

For the moment, now, let's go back to our first chapter and re-examine this sense organ about which such queerly diverse conclusions have been drawn.

The hairlets which protrude into the olfactory clefts are true nerve endings. They're the only nerves in the human body which reach the open air! Between the nerves of sight and the air there is the glassy lens; between the nerves of hearing and the open there is the ear-drum; between the nerve-endings of touch—the touch-spots, pain-spots, heat-spots, cold-spots—between all these and the outside world, there are cushioning pads of skin.

If you have ever had one acting up in a tooth, you know how sensitive an exposed nerve is. You know that nothing will deaden such a nerve, except an anesthetic. Ignoring it won't do it. The most resolute insistence that there's no feeling there won't, either!

The only exposed nerves in the human body! And they're alive, well-nourished. The olfactory area is better supplied with blood than any other part of the human body.

And, as if to guard against the sense messages being too weak, there are mitral cells which, as Herrick tells us, collect the nerve impulses from more than one of the olfactory nerves, and so step up the impulse, making even the faintest sensation strong enough to reach the brain.

And finally, the brain itself. The olfactory portion of this has dwindled, we are told. Why? Because of the tremendous overgrowth of the cerebrum, the great associational area of the brain. A hundred years ago, when every bump on a man's head, it was believed, meant a region with a particular function underneath, this would have seemed important, because so much of the cerebrum was marked off as the private grounds of sight, hearing, philoprogenitiveness, and so on. Now we know that there is no such hard-and-fast division. Helen Keller, without sight or hearing, makes good use of this associational area, simply through the messages of touch and smell.

We know that we do not use one tenth of our brain-power. There is room in the cerebrum for all the associations and thoughts our senses bring us, and ten times as many more. And—every bit of evidence available tells us that this same cerebrum stores olfactory associations as well as visual, kinaesthetic, auditory and tactile ones.

A farmer, nowadays, instead of keeping his wheat in a shed at home, stores it in a government-bonded warehouse. Similarly, the olfactory tract has turned over to this newly-evolved cerebrum many of the associations it formerly, and of necessity, kept to itself. The farmer's sheds are smaller now; the olfactory tract and hippocampus are diminished—for the same reason. Farmer and olfactory region know that the products of their labor are just as safe—in public storage. And—the farmer raises just as much wheat as he ever did—and the sense of smell gets just as many sensations!

Were it not for one thing, the olfactory area could diminish still more, and become, apart from the sense-organ, as minute as the other sense-connections in the brain, those of the eye and ear, for example. That one thing is this: not only is it difficult, almost impossible to draw a line and say: "Here the sense-receptor of smell ends; here the brain begins"— *the sense of smell is a brain in itself*. It differs from all the other senses in this. For it has a cortex of its own; a nervous system of its own; effectors of its own. Later, we shall consider these.

Now, it seems well to bring up another curious fact about this sense, another property it has that none of the others possesses. And to do so, let's abandon the time-honored division of the senses into five, and consider them in the more modern way as numbering somewhere between fifteen and twenty-seven.

These Herrick divides into three classes: the proprio-, intero-, and exteroceptors. The proprioceptors are the four touch-senses: heat, cold, pain, sensitiveness; the organs of balance of the inner ear; the common chemical sense; and the kinaesthetic sense. The exteroceptors are the two senses of sight: (1) the outlines, (2) the colors of the object viewed, as determined by the cones and retinal rods of the eye; the sense of hearing; the sense of smell. The interoceptors are the four taste sensations; the sense of smell; and the following: the organs of hunger, thirst, and nausea; and the receptors of the respiratory, circulatory, and reproductive systems.

The sense of smell is the only one of the senses which is both extero- and interoceptive. That is: it is the only sense-organ which receives messages from both without and within the body.

One more fact: excepting the taste-buds, the sense of smell is the only interoceptive sense whose location has been definitely determined. The others are assumed, because of the results observed.

Dorsey, among other psychologists, does this. Dorsey, in the same book, "Why We Behave Like Human Beings," says that Science will consider psychism when Science finds an organ that receives psychic stimuli, and not before.

Logically, the same statement applies to these supposed internal sense-organs, also. Dorsey, here, is running with the hare and hunting with the hounds. He assumes the presence of these receptors because, for the purposes of behaviorism, they are required. In the matter of psychism, which has no place in the behaviorist credo, he adopts the high, noble and correct attitude of pure science.

But if we adopt the same praiseworthy attitude towards these imagined interoceptors, what then? We should believe they're there when someone proves they're there, and not before, should we not?

The human body has been carved, sliced, and dissected times without number. Even the tiny pineal gland has been known for nearly two hundred years—since Descartes claimed that in it he had at last found the seat of the human soul.

It would seem surprising that a single one of these receptors should have escaped observation. Yet we are asked at one wild jump to believe in the existence of six, not one of which can be pointed out.

We *know* the sense of smell is an interoceptor, and *we know where it is!*

The thing has significance—but we'll leave it pending.

After observation and experiment, three students of this unique organ have reached the following three conclusions:

"The sense of smell," says Hudson, "is nearly as keen in little children as in the inferior animals. It seems to diminish as we grow older, until it becomes scarcely worthy to be called a sense."

"The sense of smell," says Havelock Ellis, "which is weak in children, becomes suddenly sharpened at puberty."

"The sense of smell," says Ponder, "is the only one of the senses which grows stronger with age."

—Seems as though someone is wrong, you think?

None of them are, entirely.

Let's review here what we've determined thus far. These facts seem —to me, at least—absolute:

All things are odorous.

Man has an individual odor.

This odor may be altered by illness, environment, diet—and emotional upflarings.

There is no physiological reason why man should not be able to use his nose as an animal does. There are no signs of degeneracy in the organ itself or in the brain behind it.

Certain individuals—savage and otherwise—with normal olfactory equipment, have a sense of smell which can be favorably compared with that of lower animals.

—And yet man generally, civilized man particularly, makes no such conscious use of the sense.

Why not?

There is a taboo.

CHAPTER FOUR

THE JEWS HAVE a taboo against pork; the Catholics one against eating meat on Fridays. And laws are taboos; and all customs. But, compared with the one we consider now, all others are as atoms beside the sun. For this thing is tremendous, overwhelming.

It is older than human history; it has *made* human history what it is. War and reason and madness and dreams rest upon it; psychology fails because of it; the conscious and subconscious states of mind are its direct results.

With the recognition of it, a bitter evolutionary battle is ended. For this taboo is an acquired characteristic—definitely inherited.

The taboo against the sense of smell!

All men have it, though savages in lesser degree. Babies are born with it as an ancestral memory. Every child has it repeated to him over and over again. And—with this repetition, what was a dim memory becomes an unbreakable law of consciousness.

You must not. smell anything! You must not receive any messages from your nose! You must not believe that your nose has any thing to give you! . . . Continually, for a hundred thousand years, ever since the beginning of language!

The taboo has achieved almost a complete success as regards conscious messages from the outer world. But, succeeding so—what a mess it has made of the human mind!

For to this good day the sense of smell does not know that it is tabooed. Twenty-four hours a day it works, sending messages into a brain which while consciousness lasts rejects them . . . Every one of them it can.

Ars est celare artem. Here is art in perfection, for the taboo has become so much a part of human consciousness that the students of mind —the psychologists—have not even realized its existence.

And so—what a mess it has made of psychology! a science in which the laws change every ten years; in which every man is his own Justinian; in which each disciple breaks away to form his own school, "Appropriating," as Parker says, "every word in the English language to do it with."

Try to imagine an astronomy built around an unquestioned belief that the world is flat. How far would it get in reconciling the physical facts with this belief? What wild theories would be advanced to explain them! How diverse the theories would be!

—As diverse as those in present-day psychology.

It is extraordinary that the mind-students should have ignored the existence of this taboo. For here is certainly a thing as plain as the nose on your face. You—or a visitor from Mars—ignorant of the function and power of each sense, but aware of the sensitiveness of nerves, would inevitably consider this sense, with its exposed nerve-endings, infinitely more powerful than any of the others. You would think that in the walking chemical laboratory that is the human body, a free chemical sense would be of paramount importance. And when you learned that the sense *can* function, but apparently does not, you would insist that something prevents it from doing so. And—we go back to our automobile analogy—what *can* stop it, except a writ of prohibition? In other words—a taboo!

And the exceptions you've already thought of are no disproof of this taboo's existence. Somehow they've escaped it, that is all. True taste remains free. So also does an external group: every other true odor that still reaches consciousness. There are surprisingly few.

We recognize these survivors here; we shall deal with them later. For each group the explanation seems adequate.

Our prime need now is knowledge of the taboo, and its cause. And to obtain this, we must go a long way back, to the invertebrate stage of existence, when the flatworm was the highest form of living creature. This worm was much like the annelids alive today. It had no brain, no spinal column, nothing but two chains of nerve-ganglia running through the back of its segments and joining at the head to form a larger ganglion. Here, the nerve lines made a sort of cortex with the nerve-endings of the worm's only distance receptor, its only means of communication with the outside world, a sense of smell. The creature had no sense of hearing, no vision; nothing but smell and a cortex, the nose-brain, behind it. Yet it managed to cope with the outside world— and survive.

In every living creature above this worm, no matter how high on the tree of life, we find this same arrangement. As the eons pass, the creature develops a backbone and a cable of nerves inside it. It gains a sense of sight, a sense of hearing, four legs. Finally, it stands upright on two of them. But always, *always,* there remain the double chain of ganglia, the nose-brain, and the sense of smell. It is as if Nature has always been afraid that her new inventions, backbone, brain, and spinal cord, will turn out unsuccessfully at the last, and life will have to revert to its flatworm stage.

The double chain of ganglia, the nose-brain, and the sense of smell make up an ancient system. In man and other higher animals, the senses of sight and hearing, the thalamus (the switchboard through which these sense-messages pass to the brain), and the spinal cord comprise another and later one.

There are connections between the two systems. Messages may be sent back and forth. Between the old sense and the new ones are interplay and co-operation, often. And there is the cerebral cortex to act as a common storehouse and distributing point for the faculties all of the senses have acquired. But—when all is said and done, there are jealousies and disharmonies, too.

The master of the house prefers his new mistresses to his old one. In fact, the new ones have sent the old queen to the kitchen, and the master, who likes a proper dinner, is well content. But the old servants have not forgotten her. They grew old obeying her, and to them she is the mistress still. Neither the master nor his young mistresses know how to handle them. New servants are not to be had. And so the master is frequently surprised and dismayed by what happens in his own dwelling, for he has no idea that the old gray lady in the kitchen still rules a house in which long ago she queened it alone.

The taboo, I think, began at the very beginning of articulate speech, man's last acquisition. Before, the senses worked well together, and man recognized friend or foe by what the three great senses told him. Sight informed him that here was a person or thing, familiar or not, moving or still. Sound signified movement always, familiar or not, somewhere. It was upon his sense of smell that man relied in the last analysis. For his nose told him of the things that meant life and food and danger and the sweet fragrance of sex, and according to the message this sense brought, he ate or ran or fought or made love. Smell was the oldest, the surest, the one he put his faith in above all. The other senses were simply auxiliary. It was perhaps the ideal arrangement— and I doubt very much if man had a mind divided between the Conscious and the Subconscious, or had psychoses, or any need of psychoanalysis, then.

Speech changed it all, for speech cut away the ground on which man's whole life-structure was built—his belief in the accuracy of his sense of smell.

Somewhere between a hundred thousand and a million years ago this occurred. Man's oldest and most highly developed sense met man's newest and most rudimentary acquisition, language, and conflict ensued, because the sense was too highly developed for man's limited vocabulary.

Apocryphally I tell what happened. At the moment First Man is already gregarious, already adapted to family life. He and his woman sit in the fork of a tree. A second woman—young, red-headed, lately widowed, and full of IT and the joie de vivre—comes swinging through the branches towards them. First Man, who is old enough to be a connoisseur in such matters, sees her; notices her red hair, her graceful movements, that subtle something about her. Perhaps he sighs a little, regretfully. And then—he turns to his wife and tries out the new human acquisition, language, confidently, proudly.

"A woman!" says First Man.

"Yeah," says his wife.

"A red-headed woman," adds First Man.

"Probably dyed!" wasn't in the language in those days, so his wife can only repeat her assent.

"Singing," says First Man.

His wife grunts.

And—"*Smells sweet!*" says First Man.

"*Stinks!*" says his wife.

Probably because even then a man understood that no wife understands her husband, First Man simply—and hurriedly—crawls down from the tree and goes somewhere. His confidence in his own sense of smell remains unshaken.

He meets his best friend, and the two sniff in amicable recognition. They sit down.

First Man notices a bush nearby on which are clusters of dark-colored berries. It so happens that such berries have been ripe for a week, but First Man, who has been on a hunt somewhere, does not know this. Nor does he know that the tribesmen have eaten too many, and that most of his fellows have been suffering from indigestion as the result.

First Man eyes the berry bush.

"Berries," he says.

"Yeah," his friend assents.

"Black berries," First Man specifies.

"Yeah."

"Smells good," says First Man, confidently.

"Smells *bad!*" his friend insists.

Once more First Man is puzzled. Perhaps he consoles himself with man's peculiar consolation: a proverb . . . Maybe he was a great man—and invented the proverb.

"One man's meat," he remarks, "is another man's poison."

And again he wanders off.

But that very day the debacle ensues. First Man and a companion have gone into the woods together. His partner, we'll say, is of different race, captured and adopted into the tribe while young. The two carry clubs.

They meet a stranger, similarly armed. The stranger, it happens, is of the outland tribe that First Man's companion was born into.

"A man!" says First Man.

"Yeah," his friend nods.

"With a club."

"Yeah."

"Enemy smell," says First Man, preparing for war.

"No!" says his companion. *"Friend* smell!"

It must have been cataclysmic—this realization that the same object could have a different smell to each man. With no thought of blasphemy, I think First Man must have felt somewhat like Christ on Calvary: "My God, my God, why hast thou forsaken me?"

Here was the sense that more than any other had ordered his whole life, and now, with the birth of language, he found that he could not depend on it. He felt abandoned, lonesome, *lost!* And when he found that his companion heard ear to ear with himself, and saw eye to eye, and had the same sense of touch—First Man lost all confidence in the ancient sense. He was afraid to mention what he smelled, for fear of being laughed at. When he *did* mention it, he met, often, a sharp difference of opinion. He wasn't sure whether he was right or not.

And he damned the sense which had thrown him into such confusion!

But why did the same thing have different odors to different men? Why does it continue to do so—even to this day? Why is it in this strange way unlike the other senses?

Here, I think, is the missing bit of the jigsaw puzzle, the small discovery that justifies this book. It is the one thing that has never been taken into consideration, the real reason why the "island" is "untouched by Science."

Remember, in our second chapter, what was said about man's personal odor? How it walks with him; stops with him; sleeps with him; accompanies him always, as much a part of him as his past?

This is the result thereof: *Man smells his own individual odor all the time.* Any outside scent that comes to his nostrils is modified by this. In other words: what man smells is a resultant of an odor from an objective stimulus, and one that comes to him subjectively, from himself. And this resultant, and not any one of its components, gives rise to the reaction.

It seems to me that this thing should need no proof, because factually it is so. Admittedly two smells blend to form an entirely different scent. Here, we have two odors, always. The act of inhalation (or—in the case of smell's taste-function—of exhalation) automatically causes the blending. It's as simple and certain as arithmetic.

Perhaps you concede the fact, but deny its importance on the ground that I am operating like the old Negro who, when accused of adulterating the rabbit sausage he had for sale, insisted that it was fifty-fifty rabbit sausage.

"And what do you mean by fifty-fifty?" he was asked.

"One rabbit and one ox," he said.

There is no such disparity here. Both external and personal odors at all times must be appreciable components of the affecting odor which is the resultant of the two. Relatively they may and do vary, and the resultant, consequently, depends upon their respective strengths. The internal, personal odor component may be overwhelming, as in illness. Then the sufferer loses his sense of smell and "taste." A cold will have this effect. Indigestion will make a man conscious of his own breath and of nothing else.

Any doctor will admit that a man sickening for a grave illness knows that there is something wrong days before the doctor himself can determine it. The sick man shows no symptoms of any specific disease, but he feels stopped-up, miserable, uninterested in the world. Actually, he *smells* his approaching illness, to the practical exclusion of outside odors.

There are times too when the external stimulus is the more important component. In taste, this is certainly so. The attitude of attention seems to permit a man to concentrate definitely on some external object —but even in this case personal odor remains an affective factor.

If we probe evolutionary history, we find the explanation of this. The sense of smell had its beginnings when there was no other interoceptor and no other distance receptor. It had to do all the work, and the one successful way to do it was by adopting a strictly utilitarian attitude toward everything. Instead of asking, vaguely: "What does this new stimulus mean?" its question was always a definite: "What does this new stimulus mean—*to me—as I am—at the moment?*"

External and internal odor-messages fused to form the one answer, cause the one reaction.

Even today, smell phrases its questions thus. The other senses don't. They can't! For the nose remains the unique intero-*and*-exteroceptor.

The eye and ear, however, reveal somewhat comparable organic conditions. There are always inter-aural noises (caused by the pumping of blood, etc.), which the sense of hearing must receive in combination with outside tones. Because of this merging, the resultant sound pattern is modified. In this case, of course, the modification is very slight —but it occurs.

The eye is a distance receptor pure and simple, yet the internal-external blending takes place here, too, for there is always idio-retinal light, emitted by the eye itself.

Again, admittedly, the modification is slight. However, the presence of these combinations in the other two distance-receptors proves that there is nothing structurally antagonistic in human anatomy to our smell-combination idea.

It is surprising to find these resemblances in the other distance organs, for they developed eons later, and have been exteroceptors from the beginning ... Perhaps Nature followed, insofar as she could, a foundation plan which had already pleased her by its success.

In the fact that the individual's reactions to olfactory stimuli are modified always by his own varying odor lies the answer to many of the riddles of Osmics. Because of it, man has entirely different reactions at different times to the same object-stimulus. Because of it, man lost all faith in the sense on which he had depended most. Because of it he fought Nature; did violence to his own mind, and made his taboo.

Civilization today, with a dictionary of three-quarters of a million words, has no vocabulary of smell. Even the verb "to smell" is difficult of use because it has both the active and passive meanings. Over and over again the writers on Osmics lament the inadequacy of language. Such words of scent as we have are simple object-words, which in use have the connotation of sight much more than of odor. Such words as rose, goat, garlic, give an image of the object even when used as descriptions of the scent.

It is difficult to see how this could be otherwise. Language is descriptive imagery. For so many years it has had its development under control of the sense of sight that now every word brings up vision. And if, with hundreds of thousands of English words, with Greek and Latin and German and French roots to draw from to build new words wherever needed—even now we can not make a vocabulary of smell, what could primitive man do with his three or four hundred nouns and verbs a million years ago?

Nothing! He couldn't even discuss the question of differences. There were no gradations between yes and no, good and bad, safe and dangerous. In the matter of this sense, with its infinite shadings and its infinite swift changes, he was inarticulate.

To him, as to Science today, the thing was inexplicable. There was something terrifying about it.

In time, this mystery of mysteries took on an animistic tinge. Man found that he developed abnormal perceptions, abnormal sensitiveness, under certain conditions—and these conditions were always basically concerned with the sense of smell. Fasting, incense, the breathing of strange odors and gaseous emanations rising from profound chasms—all had impressive results.

Fasting diminished personal odor. There was no internal smell of food, less smell of bodily wastes. Outside stimuli strengthened, because more and more they became a greater percentage of the actual odor a man smelled.

And so with the incense, and the burning myrtle leaves, and the earth-risen emanations. They had strange effects, and gave rise to strange sensations. The primitives made these things messages from the gods.

For the taboo was part of man's life by now!

As civilization grew, as language gained in scope and imagery, the taboo continued ever stronger. By the time man attained intelligence and reasoning power enough to consider the matter, there was nothing in man's consciousness to consider. The taboo had become an acquired characteristic.

Because civilization and language advanced this taboo, conversely: the lower the civilization, the smaller the vocabulary—the less power the taboo attained. And so: certain savage tribes have the olfactory sense—almost—of lower animals.

... The Taboo! A tremendous thing, and tremendously difficult to create—because there was no co-operation. The sense of smell kept right on receiving stimuli, transmitting messages to the brain, being fed so that it could do its work. For man has an old, old body compared to that young brain of his, and the oldest parts of this body act in a way of their own. They work *automatically*—as the sense of smell has done through all the hundred thousand years of disgrace.

Over his new organs; over his new brain; over the conditioned re-flexes he learns each lifetime, man has almost absolute control. Nor-mally, he has absolute. But—and here is an interesting thing—in times which are not normal he has no control at all. For his old body, and his old organs, and his old sense take charge—and these are in such times uncontrollable. Those parts of man's new humanness which can-not work automatically are not allowed to work at all. They are stopped by a certain-sure method: they are not given raw materials to work with. The old body, in good times a generous host, turns quite other-wise in times of stress. It is interested in itself only; in its safety, in its food, in its continued existence—and that's all! The new brain is left clinging desperately to its taboo—while the body acts as if there were no such thing.

This one fact finally: the taboo is part of consciousness. The sense of smell is part of the living body. The conflict is still on, and each generation must re-learn the lesson that is so easy to forget.

In the learning is lost the little consciousness of smell ancestral inter-diction has left us. What was it Hudson said? "The sense of smell, which seems to diminish as we grow older until it becomes scarcely worthy to be called a sense, is nearly as keen in little children as in the lower animals."

The taboo must be re-learned!

We take up now a notable exception to the proscription: taste. It had survival value; but so did other odors that lost out. Its salvation really came from the fact that food, once in primitive man's mouth, was nobody else's business. If he liked it, it was good; if he did not like it, it was bad. No one was likely to dispute his opinion but the cook; and if a man beat his wife because she served him poor victuals and then lied about it, the rest of the tribe admired him, pitied him, and pointed out what had happened to the woman as an object-lesson to their own wives.

A second, supplementary reason for continuing, conscious taste is all the more interesting because it seems evidence of the taboo's reality, proof that a prohibition, and not inability, stops one from using his nose.

Man and a lower animal can not marry because such mating is an im-possibility. In the South, a white woman and a man with any Negro blood whatever cannot marry because there is a social and legal taboo against it.

Some quadroons look like sunburned white men. Such a fair-skinned Negro, keeping his mouth shut about the tarbrush in his ancestry, can go down South and marry a white woman without difficulty. The taboo will not be raised against him.

Why not?

Because no one knows that the taboo applies!

Primitive man and his descendants placed a taboo on the sense of smell. But—primitive man did not know that real taste is a function of this sense. And so—we taste our foods. The taboo was not raised against taste.

Why not?

Because no one knew that the taboo applied!

What other possible explanation is there of this anomaly—this putting a "No Visitors!" sign on the front door, and a "Welcome!" on the back door of the house of smell!

CHAPTER FIVE

IT IS MAN'S WAY, in investigating any object, to bring every sense he can to his assistance. Thus it is usually possible to attribute to a favored sense more than that one actually does, to the discredit of another sense. Under the taboo's guidance, this is continually being done. Messages received by the nose are ascribed to the eye or the ear, so that we say: "We see" what we really *smell*.

In normal folk, it is hard to prove the facts otherwise. But, because all men are not born perfect, it happens that we can consider cases in which it is impossible to give other senses undue credit. In these cases, the taboo is re-learned with the greatest of difficulty, and never so well. For the handicapped unfortunate faces facts and must attempt to explain them in a permitted manner, though debarred from the normal easy solution: the attributing of these cerebral messages to another and utterly wrong sense-origin. Some of these attempts drop to manifest absurdity.

For a proper example we contemplate a sadly misused lady and a woman who with the best intentions in the world, perhaps—played hell! I speak of Helen Keller and her teacher, Mrs. Sullivan.

Miss Keller in early childhood lost her hearing and her sight. It is entirely probable that had she been left to herself, she would have remained reasonably happy, and utterly unaware that she was in any way more unfortunate than other humans. Such vague memories as remained to her would have held as little meaning and truth-seeming as the memories of any other adult concerning his earliest infancy.

No child of two years has a concept of the senses, or of their use. Before Howe of Boston, *and* his daughters, *and* some several diary-keeping lady teachers all found that publicity and encomiums awaited teachers of the blind-deaf, such unfortunates were let alone, and from all accounts lived happy, harmless lives in whatever curious world it is that Nature makes for them.

But Howe had gained fame, *and* Julia Ward for a wife, *and* three biography-writing daughters—simply because he had supervised the instruction of Laura Bridgman.

Here in Helen Keller was another deaf-blind child to be taught— and ensuing rewards to be reaped! Mrs. Sullivan made the girl her life-work.

To what result? At their first meeting, Helen had two senses left— touch and smell. Mrs. Sullivan took away one of them—and the world praised her noble work!

I confess I can not write dispassionately about this. It seems to me too much like taking away "from him that hath not, even that little that he hath." There is such an instinctive rightness about what we learn from Miss Keller's own confession: that at the beginning she hated Mrs. Sullivan. The error came later, with the dutiful love and dutifully deep gratitude she feels towards her teacher now. For Helen Keller does not know what wrong has been done her, just as she does not surely know which of her two remaining senses brings her the messages she receives.

Mrs. Sullivan, using established methods, could train Helen Keller's sense of touch. She could also teach her to talk—and this was no miracle, for Laura Bridgman very nearly taught herself to use language.

But on the debit side: Mrs. Sullivan re-established the taboo in Helen Keller's mind. There was no honor, no glory for Mrs. Sullivan if Helen Keller received messages from without through an untrained and untrainable sense of smell. There was much honor for the teacher—if she received them through a sense of touch trained by Mrs. Sullivan.

And so, perforce, Helen Keller attributes all things, dutifully, to her sense of touch. She is so pitiably anxious to be as near normal as possible. She knows—because Mrs. Sullivan told her—what Philip Curtis expressed so perfectly: that smells are regarded as sensory swear-words. A gentleman may have a nose, but no gentleman is supposed to do very much smelling.

Can't you hear that Massachusetts lady teacher signalling: "And no lady, either, Helen!"

Except, of course, sweet odors. So Miss Keller admits with relief that she can smell sweet odors: "When the book of the year opens at June, I want to drop my work, whatever it is, and enter the kingdom of delight. It is a wondrous woof of odors."

"At home I have always been where I could breathe the woodland air. Winter and spring have always brought me messages—wind-blown —across marsh, brook, and stone-walled field."

". . . My greatest pleasure is outdoors" . . . "I notice the loss of odor when climbing over the mountains in a plane" . . . "The sense of smell is the aesthetic sense, I think, even more than sight. I know that odors give me a vivid conception of my surroundings. I can smell my landscape—because when I walk or drive through the country—odors tell me of field and stream, etc. . ."

On a railroad trip she recognized certain things: a field, a tree, a fire, a building, entirely by smell.

Once, just for a moment, the taboo slipped. McKenzie told her about the Indians' ability to smell fire farther than the eye could see it. Helen answered quickly: "I also can smell at a vast distance."

But nowhere in her writings does she mention a single scent that it would be unladylike to smell. Often, beneath her words, we can see the struggle:

"The senses assist and reinforce each other to such an extent that I am not sure whether touch or smell tells me the most about the world. Everywhere the *river of touch* is joined by the *brooks of odor-perception.*"

When brooks of water join, the river thus formed is water, too.

"Each season," says Miss Keller, "has its distinctive odors. In autumn soft alluring scents fill the air, floating from thicket, grass, flower and tree—and they tell me of time and change— of death and life's renewal —desire and its fulfillment." . . . "I tread the solid earth—I breathe the scented air. Out of these two experiences I form numberless associations and correspondences. Blindness has no limiting effect on mental vision."

—*Scented air, sweet flower smell*—and that is all! What about contacts with humanity, sensations of humanity?

Here is where Mrs. Sullivan's work comes in.

"It is all," says Helen Keller, "from my sense of touch." She feels persons away from her by their vibrations.

Now, it is true that Dr. Tilney discovered that, tested objectively, Miss Keller has no keener sense of smell than ordinary humans, and equally true that in her sensitive finger-tips she has a much more acute (or at least, well-trained) sense of touch than most persons possess, but we ought to get to the very bottom of the matter. As compared, one with the other, which is the more sensitive in its very nature—smell or touch?

At first thought, any weighing of the respective powers of the two seems impossible—but a curious physiological discovery gives us an actual basis for comparison.

It was found that when the nerves leading from the taste-buds of certain animals—rabbits, for instance—were severed, the taste-buds disappeared shortly. All that was left was an ordinary sensitive touchspot. When the nerves grew back, and innervated the spot again, the taste-buds developed once more. Therefore, it is supposed, the taste-buds are specially designed touch-receptors, specially selective for certain chemicals. It is presumed that they are part of the sense of touch, as the pain-spots and hot- and cold-spots are, but by all odds they are very much more sensitive.

The taste-buds are more sensitive than any other form of the sense of touch—and we can compare the taste-buds with the sense of smell.

The bitterest known substance is a strychnine compound. The most odorous substance is mercaptan. It we take the weakest solution of strychnine recognizable as a taste, and compare it with the weakest solution of mercaptan recognizably an odor, we find that the mercaptan need be only one millionth as strong. The sense of smell, thus compared with the taste-buds, is one million times as acute.

A very few things affect both smell and taste-buds. Ethyl alcohol is one of the few. Tests with ethyl alcohol establish the fact that one twenty-four thousandth as little as is needed to cause a perceptible taste will make a recognizable odor. In other words, fairly tested with the same substance, smell is 24,000 times as strong as taste.

If the taste-buds are more sensitive than any other touch spots, and smell is 24,000 times as discriminating as these taste-buds, then smell is *more than* 24,000 times as discriminating as touch!

Yet Helen Keller attributes all sensations except sweet odors to touch.

In music, certainly the vibrations to which humanity is most sensitive, Miss Keller needs a tremendously powerful instrument, a piano or an organ, of which even ordinary persons can feel the throbbing. With her fingers on one of these, she gets her reactions. And yet, with all her vaunted sense of touch, she has never been able to distinguish one tune from another.

When she writes of her contacts with the outside world, however, she attributes them wholly to touch. I quote some of these sensations, these contacts: "I have endlessly varied, instructive contacts with all the world, with life, with the atmosphere, whose radiant activity enfolds us all. The thrilling energy of the all-encasing air is warm and rapturous. Heat waves and sound waves play upon my face in infinite variety and combinations until I am able to surmise what must be the myriad sounds that my senseless ear has not heard. The air varies in different regions, at different seasons of the year, and even different hours of the day."

Touch?

"The silence and darkness which are said to shut me in—open my door most hospitably to countless sensations that distract, inform, admonish and amuse."

Touch?

"Often footsteps reveal in some measure the character and the mood

of the walker. I feel in them firmness and indecision, hurry and deliberation—activity and laziness, fatigue, carelessness, timidity, anger and
sorrow. I am most conscious of these moods and traits in persons with
whom I am familiar."

Touch?

Never a word about odor, never a specialized instance of human-smell
reaching her nostrils. And yet—Helen says that as a child, it was two
years before she could believe that dogs didn't have personalities like
her own. They seemed to have, to her.

And did she get this from her sense of touch? Of course not! It occurred before the advent of Mrs. Sullivan—and the taboo.

Laura Bridgman was also a deaf-blind woman—but in this case not
so well-disciplined, perhaps. For Laura Bridgman was also conscious
of moods and traits in persons with whom she was familiar—and Laura
attributed it to her sense of smell. She knew all her acquaintances by
smell; and she could sort linen, after it came from the wash, by the odor
alone!

There was James Mitchell—a deaf-blind boy of whom Dugald
Stewart wrote. James was not trained at all—but he was always able
to tell when a stranger entered the room, and to give an opinion of his
character, evidently (says Stewart) through smell. And, by his nose, he
formed a concept of death, for he was found, the day after his father
died, sobbing bitterly on the grave.

He was deaf and blind, remember, and this was before Braille, before touch-training of any sort. Yet he knew emotion; had contacts with
the world outside—for the ancient ban was but a weak, inherited memory. No one had ever told him it was ungentlemanly to smell. No one
had ever told him he *could* not smell. No lady teacher—had re-taught
him the taboo.

And so, he could judge character through the use of his nose, *consciously,* as a dog does every day.

Poor Helen!

Palliation, if any, for her treatment by society, as embodied in Mrs.
Sullivan, would have been on the grounds of the greatest good to the
greatest number. Civilization had need of what Miss Keller could have
told it about the sense of smell. Because of the benefit this information
would have been to humanity—and for no other reason, the injustice
to her might have been condoned.

It *was* injustice. For Helen Keller was ignorant—and satisfied. Civilization showed her a new heaven and a new earth—and told her it was not for her; showed her a promised land she could never enter; condemned her, her life through, to long for impossibilities; to struggle bravely against an unyielding fate.

To nobody's advantage—*because the taboo was taught her, first!*

CHAPTER SIX

LAURA BRIDGMAN, it was said: "could sort linen after it came from the wash, by the odor alone."

Obviously, touch and the kinaesthetic sense were employed by her, also. She could not sort the linen without handling it; she could not handle it without moving her arms. But—no matter what other senses were necessary aids to the accomplishment of the work, smell was the essential, directing force.

Recurring in one aspect or another, this co-working of the senses will be a spectre to haunt us throughout the remainder of this book. Even in situations in which man's responses are primarily external-odor ones, eye and ear may play some small supporting part, and the problem will be whether we should stop, each time, to give them their microscopic meed of credit.

Inevitably, by attributing its true importance to the sense of smell, I shall detract greatly from other distance senses. But I do not wish to reduce this argument to an absurdity by claiming more than the nose is entitled to, especially since much of our own reasoning draws force from the absurdities patent in the present custom of giving smell no credit at all.

Common knowledge and belief, heretofore unchallenged, have bestowed on eye and ear an unnatural monopoly. This I would break down. But I have no desire to leave any consequent impression that all men are both deaf and blind.

I have been forewarned that there is danger of this—and I distrust my own eagerness. And so, here and now, to save unending qualification throughout the remainder of this book, I explain that when I state: "The sense of smell is . . . ," I do not mean that it exclusively, unassistedly is . . . I mean simply that no matter how many others are adjuvant, smell is the sense that gives the "Go" signal, the sense in actual command.

All you have read heretofore has been in the nature of an introduction. The sense of smell has been so entirely discredited that I was afraid any claims I made for it would seem ridiculous. And so I attempted to set down data more or less unknown simply to put you in a receptive state of mind. It is my hope that you have gained a measure of respect for the nose on your face; that you are puzzled why Science has ignored it so long; and that you wonder perhaps if Science in so ignoring it has not unconsciously created riddles it now finds insolvable.

Part of the way, at least, we should be in agreement. These things, surely, are acceptable:

1. Man has an individual odor.

2. This odor, a resultant of the bodily chemistry, varies with any chemical change in the body.

3. The human olfactory organ, consisting of living nerves exposed to the open air, is potentially the most sensitive receptor in the human body.

4. There is no factual basis for the assertion that man has lost his sense of smell. On the contrary, there are numerous instances of men with an odor-sense fairly comparable with that of lower animals.

Since all our knowledge leads us to the belief that civilized, normal man *can* smell but does not, we find that we are faced with a taboo, for a taboo by simple definition is: what man can, but does not permit himself to do. Whether the version given of the genesis of this age-old ban be true or false is of little importance, so long as the taboo itself is admitted.

Last of all, I advanced the one fact which has never been considered, which is actually the answer to a whole chain of riddles, this: that man smells his own odor, and that thereby every stimulus reaching his nose comes to it as a composite scent, of which his own variable aura is an integral part.

On that one assertion our complete theory rests. It is the crux of the matter; the nucleus which animates the whole. Admit it—and quite likely we shall be together all the rest of the way; deny it—perhaps you'll turn believer, later on.

It seems time now for a basic hypothesis. You will note, on reading it, that the two groups of conscious odors are not mentioned. The first: taste-messages, we have already discussed and claimed as a specific proof of the taboo. The second group comprises all the various smells, perfumes, stenches, consciousness knows. These form an objection to our theory that must be answered fully and satisfactorily. In the next chapter, I shall attempt this answer. Until then, I ask you to reserve judgment.

There is a law—the Law of Parsimony—by which every hypothesis stands or falls. Watson phrases it thus: "If an explanation is simple, straightforward, adequate, it suffices."

From a general text-book: "Psychology," by Perrin and Klein, used at Yale University, I take the following: "The Law of Parsimony: The principle of selecting an explanation or theory that makes the fewest necessary assumptions to cover the facts is usually referred to as the *Law of Parsimony*. . . . The psychologist is governed by this law in interpreting the phenomena of mental life."

Our assumptions total: *one* . . . that the sole evidence of degeneracy in the human sense of smell is the common and conscious knowledge thereof.

. . . To which should be appended certain truisms: that a chemical reaction is a chemical reaction; that nerves act like nerves; and that one and one make two.

Here is our hypothesis:

1. *Man has a functioning sense of smell, acted upon by combinations of objective and subjective odors.*

2. *Smell messages, consciously tabooed, are subconsciously accepted.*

3. *The total of these acceptances is the Subconscious.*

And:

4. *The automatic actions of the body are smell-responses.*

In more detail:

1. *Man has a functioning sense of smell, acted upon by combinations of objective and subjective odors.*

By the expression, "a functioning sense of smell," we mean a sense as important to man as the same sense is to a dog. We mean, to mince no words, a sense more important to man than his sense of sight.*

"Acted upon by combinations of objective and subjective odors." Varying odors of varying relative importance. Man has a natural odor —an ideal odor, in a way. This is what his scent would be were he in perfect health. It is doubtful whether he ever quite attains this. But, from an approximation of it to an utterly different disease odor, there is a whole gamut of changes. Likewise, from a state of vague happiness, the result of this ideal odor, to a feeling of imminent dissolution,

---

* More than any other in this book, I have been urged to soften this statement. I will not because I cannot. A blind man keeps his other faculties, and lives his threescore-and-ten. But, bearing in mind the body's delicate balance, I do not see how a true anosmic (i.e., one inwardly so, too,) could stay sane for an hour, or survive for a day. (1960.)

the reaction from the disease smell—there are equally as many changes. The objective odor is fused with this personal odor; the natural reaction from such objective stimulus is strengthened, weakened, or altered in other ways by the reaction from this personal odor. The same objective stimulus, thus, may have an entirely different effect on two individuals, and a different one on the same man at different times.

2. *Smell messages, consciously tabooed, are subconsciously accepted.*

We include here even infinitely slight odors reaching the olfactory nerve-endings. Because of the delicacy of these receptors, and the step-up system of mitral cells, by which individual nerve-impulses too slight to affect the brain are joined to make a strong and sufficient impulse, we think that odors inconceivably weak, and much below the threshold of conscious smell suffice to cause a reaction. And, because it seems definite that a sensory impulse is never lost between the two (peripheral and cerebral) ends of an efferent nerve, these stimuli must cause cerebral action. Since we know that there is no such cerebral action in consciousness, the cerebration must occur in the subconscious mind.

... I do not ignore here the fact that lack of conscious cerebration is the one great argument in favor of man's loss of smell, but—if we admit that the sense of smell is a chemical sense (and this is admitted by science); if we admit the potential sensitiveness of the olfactory receptor (and all our knowledge of nerves corroborates this)—we must admit that chemical action on a chemical sense must send message-bearing nerve impulses to the brain.

Looked at thus, it seems more logical to deny consciousness of the impulse rather than to deny the actual existence of the impulse. Conscious knowledge can not tell us that there is no impulse—simply that there is no consciousness of such an impulse. Conscious knowledge does not tell us that there is no life on the planet Mars—simply that we have no conscious knowledge of such. Even this comparison understates the case. We have no knowledge of any life whatever on the red planet. We do know that other cerebral actions occur of which we have no consciousness ... It is as if we admitted that there is primate life on Mars, but denied the existence of gibbons.

The taboo adequately explains the relegation of smell messages to the Subconscious. There is always a bar between mind-divisions, whether such separation be made between the Conscious and the Subconscious, or the Conscious and the Unconscious. Freud's Censor is essentially a taboo, full-fledged.

We have physical grounds for a belief in the existence of smell-impulses. We have a place to put them—the Subconscious. We have a world of odors to act as stimuli.

And against all this, we have a conscious denial, based on nothing more than a taboo, of a particular bit of subconscious cerebration. And what is this denial, in essence, but the indignant denial—as described by Freud—by a neurotic of the possession of the Libido that is responsible for his neurosis.

3. *The total of these acceptances is the Subconscious.*

. . . If I were to change this statement to: "a large part of the Subconscious," it would be received with more credence. And yet—I will not. Not at present, anyway. There are certain instances difficult of explanation; certain situations in which eye or ear seem the receptor for messages consciousness does not receive. But I feel that our knowledge, and not our theory, is at fault. I can see no reason for inhibiting any message from eye or ear, and it seems to me that by its very nature, the Subconscious must be the result of inhibition; that always a taboo, like Freud's Censor, must keep the threshold of consciousness.

Of this I am sure at any rate: that our definition of the mind beneath will more closely approximate the complete whole than either Freud's purely sexual one or Watson's theory of instinctive behavior.

I admit the possibility of our definition not being all-inclusive, but it at least includes all of Freud's and Watson's—and more.

This is also true: Freud puts the formation of the Unconscious between birth and five years of age; Watson from before birth to three or four. Both statements are theories and will remain so. For Freud's boasted cures have the fatal weakness of being explicable in another much simpler way; and no child has any memory to corroborate Watson. There can never be any introspective proof of Behaviorism—and perhaps because of this, Watson wants to discard all introspection as worthless.

It isn't worthless—except as a proof of his theory. As Herrick says: "Watson wants to sink the ship to drown the rats."

Freud, out of hidden memories which are actually conscious ones, digs up a memory he can call a Libido—and stops.

Watson, without proof of any sort, sees effects and finds causes to match them; Freud sees one facet of the total effect, and out of conscious memories digs one cause to fit it. Watson ignores all things that weaken his theory; Freud makes one thing so important that all the others are minimized.

Our smell theory at least can be tested. It offers predictable results. If it is wrong, it can be proved so.

4. *The automatic actions of the body are smell-responses.*

... Once more I make a comprehensive claim, for by automatic actions I mean all visceral and endocrinal ones. And since so far as it goes, I accept the James-Lange theory of emotions as visceral change, I say that emotions are smell-responses. Finally, passing over the very minor role played by the taste-buds, I assert that the sense of smell is the *only* interoceptor!

I have made general statements—and with intention. I speak of the complete human mind, the entire human body, *all* the Subconscious, *all* the emotions, the *only* interoceptor. I do so in the face of certain minor exceptions troublesome to explain: Humoresque played on a violin, for instance ... and the tears that fall in a motion-picture theatre.

It would be far easier to say: *most* of the Subconscious; *most* of the emotions. I do not do so because I do not wish to leave myself any storm-cellars. If the smell-hypothesis is wrong, it can be ignored, with no damage done. If it is correct—the thing is dynamite!

For it would have its blasting effect on every science pertaining to human mind and human body; it would destroy the Augean stables of Psychology; reveal for the first time the buried bases of human behavior.

To meet the standards set by the Law of Parsimony, a theory must be simple, straightforward, adequate. This one, with its four explicit statements, no exceptions, is simple, straightforward enough, surely. Its adequacy remains to be determined.

I submit that if it can be used intact in every study of human mind and human body; if it simplifies riddles science has never solved; if it at no time reaches the dead end of the inexplicable—it is adequate.

All I ask, so far, is that you admit that there is no physical impossibility to estop it.

The real test lies in the effect. And to find out this we shall consider, one by one, the following: instincts; emotions; crowd psychology; sudden antipathies and friendships; personality; love; sex; mourning;

moods; day dreams; dreams; hypnotism; psycho-analysis; Jung's introvert and extrovert theory; Adler's inferiority complex; ghosts; haunted
houses; telepathy; clairvoyance; insanity; the Subconscious.

And—if the hypothesis holds all the way through, what is it?
The truth!

CHAPTER SEVEN

So FAR, we have almost ignored the subject of sweet perfumes. Mention was made of the two hundred *odeurs* between which a "nose" can distinguish. Helen Keller told us of her delight in scented gardens. . . . And that is all.

If you have read other writings on smell, you have been struck by this, for all other authors fill their pages with accounts of strange and curious fragrances: of the perfume jar in the tomb of an Egyptian king which, after four thousand years, still gave out a perceptible scent when archeologists discovered it; of the book in an English library on which musk was accidentally spilled while the volume was being brought from Spain, and which now, hundreds of years later, makes the room in which it stands still fragrant; of the mosque in Africa whose builders mixed musk with the plaster that holds the stones together whereby even now in the sunlight true believers get a sweet reminder of the musk maidens awaiting them in Paradise.

So many pretty stories! but to our smell-hypothesis they're merely a list of manifest exceptions that must be adequately accounted for.

Perfumes, stenches, food odors—here are external stimuli that are neither consciously tabooed nor subconsciously accepted. If there *is* an odor-ban, why aren't these forbidden? If smell has been relegated to a brain-beneath prison, how is it that these stay consciously free? Why —the problem remains—should man put a vehement taboo on one smell and not on another?

Unless the mystery can be resolved, our theory is hopelessly weakened. And, realizing this, I have asked myself these questions, many times. As a result, I have two solutions, either or both of which may be right.

Before this superfluity of riches I do not know what to do. The first explanation, which for a time I thought the only one, I offer now noncommittally. It requires a brief excursion into physiology, and a reconsideration of a century-old experiment.

The nasal septum is innervated by three cranial nerves: (1) The olfactory—really, a group of nerves; (2) The trigeminal; (3) The terminal. All three have free nerve-endings, although the olfactory possesses the only ones that reach the open air-chamber. No one knows the function of the terminal nerve. The trigeminal carries the messages of the common chemical sense, and is supposedly sensitive to irritants.

Magendie's famous experiment and the criticism that greeted it occurred nearly a hundred years ago. Unfortunately, I have read only Parker's account of the controversy. On the face of it, if this account is comprehensive, the question is still wide open.

Magendie claimed that the trigeminal was the true nerve of olfaction; the olfactory, one whose functions were wholly unknown. A dog with its olfactory nerve cut responded to acetic acid, ammonia, etc.

Critics called these things irritants.

Magendie answered that his results were not dependent upon these irritants, but could be proved by non-irritants, such as lavender oil.

His conclusions were attacked at once. In experiments, Valentin found that a normal rabbit would sniff the body of a dead one, but that a rabbit whose olfactory nerves had been cut would not thus respond.

Schiff experimented on five puppies, four with severed olfactories. The pups in which the nerves had been cut were unable to find the mother's nipples, and did not distinguish between a male and the mother, though they turned their heads away and sneezed when ammonia and ether were administered.

There was also, Parker tells us, the testimony of anosmics.

He goes on: "These and other results showed that though the trigeminal endings were concerned with the reception of irritants, true olfaction was accomplished only through the olfactory terminals, which have to do with delicate perfumes, aromas and the like, many of which are associated with food. This recognition in nasal stimulation between two classes of substances, irritants acting on trigeminal terminals, and true odors affecting the olfactory endings, is of fundamental importance, and the failure to appreciate this distinction is responsible in part at least for much of the confusion that exists concerning olfactory stimulus."

The distinction between irritants and true odors—no question but that this is of tremendous importance. The question is: *whether this distinction has ever been truly made?*

Valentin, laboring to refute Magendie, found that a normal rabbit would sniff the body of a dead one, but that a rabbit with severed olfactories would not thus respond. This proved—what? . . . That a rabbit uses its olfactory nerve in finding out about another rabbit. And this is no disproof whatever of Magendie's experiments with non-irritants—substances like lavender oil—which a normal rabbit would never encounter.

Schiff's research seems equally extraneous. It re-establishes just what Valentin proved: that an animal uses its olfactory nerve in testing its own kind.

Anosmics, we are told, denied Magendie's theory. As to this: the testimony of these supposed smell-less persons has been the preferred source-book for all scientific study of smell—and it has been throughout utterly confusing and conflicting. Using it, how far has such scientific study got?

Nowhere! "A little island . ." etc.

Anosmic testimony has never proved *anything* . . . which means that it has never disproved anything, either—including Magendie's statement that a dog with its olfactory nerve cut responded to acetic acid, ammonia, and (the thing we're especially interested in) non-irritants, such as lavender oil.

Neither did Valentin and Schiff disprove him. For suppose their vivisections were performed on a human subject.

(After all, the conclusions drawn from the experiments were applied to humans!)

The man thus tested, with his olfactory nerve cut, would not sniff the body of a dead man, and he would not—by smell—be able to recognize his mother. In neither case would his proper normal reaction have been altered by what had been done to him. For no individual today, conscious and in his right mind, sniffs corpses or shows olfactory interest in his kinfolks.

Now let's try Magendie's experiment on our human guinea-pig. With his olfactory nerve cut, as before, the victim responds to acetic acid, ammonia, and non-irritants such as lavender oil!

Precisely as a normal human, with olfactory nerve intact, does, today! Queer, isn't it!

"True odors," says Parker, "are much milder than stimulants for the trigeminal nerve, and seldom call forth strong responses."

How much response do sweet fragrances call forth?

Here's another queer thing: a comparison. A factory "nose"— expert extraordinary—can distinguish, by smell, *two* hundred perfumes. A coffee-blender can differentiate, by taste, between *seven* hundred coffees.

If these figures were reversed, they would seem less puzzling. For the sense of smell works much more efficiently through its front door than through the rear. Parker says so: "We taste relatively strong solutions, we smell much milder ones"; and our own hypothesis explains: Why?

In exhalation, two of the body odors—those from the alimentary canal and from the respiratory tract—become of major importance. There are half a dozen body odors which become part of the smell-messages in inhalation, but their total effect is never so great. And—it is when we breathe out that we taste.

Since taste-sensations must be merged with powerful body odors, they can never be as sharply clear as external smell messages, and so must be relatively more powerful than the latter to be distinguishable at all.

The perfumer uses the admittedly more efficient front door of smell; the coffee-blender uses the back door. The perfumer has the whole garden kingdom of plants to inspire him; the coffee-taster must confine himself to a single species of the madder family. And yet, by a seven-to-two margin, the coffee-taster does the more selective job.

Why?

Just another of the mysteries of smell, the pundits tell us. That is no answer at all.

This is at least a possible one:

The common chemical sense by which we react to irritants is weaker by far than even the back door of smell. If perfumes were irritants, we should have no difficulty in resolving the mystery. The coffee-taster, we should say, *smells;* the "nose" testing perfumes, is *irritated*—and because the irritant sense is so much less sensitive than the olfactory one, the coffee-taster can taste seven hundred coffees; the "nose" can be irritated by only two hundred perfumes.

But perfumes are not irritants! Stenches are irritants. Garlic-y smells, maybe—but flower scents never, you say. Listen! The Japanese, when testing perfume, get rid of smell-fatigue by sniffing vinegar.

Vinegar is dilute acetic acid—an admitted irritant. Why should acetic acid, which stimulates the trigeminal, have any effect on the smell-fatigue of the olfactory nerve? It shouldn't. But if we consider perfumes as irritants, there's no problem.

How weak *are* these delicate fragrances that come to us?

"Smell," says the Encyclopedia Americana (referring to conscious odors), "influences the respiratory processes, and the breathing of a fine odor increases the amplitude of the respiratory movements." . . . "Many odors," Havelock Ellis avers, "are nervous stimulants." . . . "Perfumes," says Redgrove, "would seem to act—not only on the olfactory nerves— but upon the whole nervous system. Their use calls for moderation and restraint—as otherwise the effect may be harmful."

Fere found that a kind of sensorial intoxication could be produced by the inhalation of odors. "It is," says Ellis, "the genuinely stimulant qualities of odorous substances which led to the widespread use of the more potent among them by ancient physicians."

And which are these more potent odors? Vanilla, musk, aromatic woods, verbena, rue, patchouli, rosemary, sulphur, saltpeter, ambergris, tobacco, lavender, civet, rose, mint, citronella, violet, menthol, camphor, clove-pink, spikenard, cinnamon, thyme, French geranium, cloves, valerian, eucalyptus—the list is limited only by space.

There's no trouble at all in finding a world-wide, age-long use of these delicate odors as medicines. There is plenty of evidence that they are drugs.

Here's Havelock Ellis' contribution: "Odors are powerful stimulants to the whole nervous system, causing like other stimulants an increase of energy which, if excessive or prolonged, leads to nervous exhaustion. In medicine, it is well recognized that the aromatics containing volatile oils: anise, cinnamon, cardamons, cloves, coriander and peppermint are anti spasmodics and anesthetics, and that they stimulate digestion, circulation, and the nervous system, in large doses producing depression."

The tests of Shields, of O. Henry and Tardiff in France, Fere's work with dynamometer and ergograph—the proof piles up that every sweet odor is a powerful drug. . . . and yet gentle ladies and gentlemen buy the stuff to the extent of several hundred million dollars annually. In this country, which for a decade denied strong men the gentle stimulation of beer!

There are odors which, if smelled long enough, will depress you so that you will commit suicide. There are others that will save you the trouble—they'll kill you themselves. Huysmans used to go on olfactory benders; so did Baudelaire.

There is a Federal Drug Act to cut down the sale of recognized stimulants and depressants like morphine and cocaine and marihuana. There are all the advertising pages in the world to increase the sale of other stimulants and depressants—but they put these in cut-glass bottles and call them sweet fragrances.

Considering "odeurs" as true odors, the question of smell fatigue is baffling. Considering them as drugs, any dope-fiend can explain it. And the fact that the sweeter the sweet fragrancy, the quicker the smell fatigue, which ought to have given the whole answer, seems to have escaped even Zwaardemaker's eagle eye.

You'll find these same sweet fragrances employed in religious festivals to turn sober folk into Bacchantes; you'll find them the first recourse of heathen medicine men who need their drug-maddening powers for prophecy; you'll find them used to exorcise demons in demoniac travesties on religion. And you'll find them, deadly demure, in ladies' boudoirs.

The Babylonians and Assyrians employed strong and penetrating odors to purify the air—and also the body—three thousand years ago. Hippocrates and Criton classified perfumes as medicinal agents. Empedocles of Agrigentum is said to have stopped the ravages of the plague in that city in the 5th century B.C. by having great fires of aromatic wood burned in the streets. During visitations of the plague in the Middle Ages, fumigations were largely depended upon to destroy the supposed "aura" or poison of the disease, generally believed to be in the air. The Great Plague in London was so controlled. It is said that during epidemics of cholera in Paris and London, working perfumers have remained immune to the disease. Richelieu was a firm believer in the revivifying force of perfumes, and during his last illness insisted on sweet-smelling powders being diffused in his room by means of bellows.

Silly old superstitions?

I quote Redgrove: "Many essential oils may be described as poisons. They have a high antiseptic value. The old time notion that sweet odors were good to ward off pestilences, illustrated by the employment at the Old Bailey of rue and rosemary as prophylactics against jail-fever, was essentially sound."

Some years ago the discovery was made that cases of tuberculosis were much less common in the flower-growing districts of France than in other parts of the country. This was attributed to the antiseptic effect of the essential oils of the plants in general. It was also noted that in the laboratories where the oils from the flowers were prepared, the majority of the workers remained remarkably free from diseases of the respiratory organs. This also, it was claimed, was due to the essential oils. The matter was carefully investigated and reports were made to the French Biological Society and the Institut Pasteur. Their observers found that micro-organisms of glanders and yellow fever were easily killed by essential oils such as cinnamon, thyme, French geranium, etc.

In later experiments, bacteria were exposed to the emanations from essential oils for various periods. In some cases, notably when cinnamon, cloves and verbena were used, the bacteria were killed in a very few minutes.

Omeltschenki, too, conducted tests on the bactericidal properties of the essential oils of flowers. He found that the bacillus of typhoid was killed in forty-five seconds in air impregnated with the vapour of cinnamon or valerian; and that the bacillus of tuberculosis was destroyed in twenty-three hours by oil of cinnamon, and in twelve hours by oil of lavender or oil of eucalyptus.

If you use a perfume as an antiseptic, it's quite all right—only there are cheaper ones. As drugs or medicine, it's all right, too—only a doctor's prescription should accompany. But—if you use it in exterior decoration, be careful! Some forms of the stuff will kill a rabbit or frog, quite quickly.

So now, how strong are these sweet-scented, so-called true odors?

Pretty powerful, are they not?

And Parker makes strength a distinguishing characteristic! In his own words: "Irritants or stimuli for the trigeminal nerve induce violent reflexes, respiratory and the like. True odors are much milder and seldom call forth strong responses."

Let your dog smell your perfume bottle and see what he thinks! He'll sneeze, which is one of the most violent of all respiratory responses.

Looked at in this way, we can divide true odors—*conscious* odors, rather—into two classes: trigeminal stimuli and food odors. That is all. The smells of food—good, bad, and indifferent, ready to eat and rotten—are true odors, the only conscious ones.

The division follows the observed effect of the two classes. The trigeminal stimuli—irritants *and* perfumes—affect the respiration, the heart, and the blood. Food odors stimulate the salivary and other food-conditioning glands. And—food odors cause the first processes of digestion, an automatic function over which consciousness has no control at all.

As I stated, the above explanation of certain conscious odors is definitely open to doubt. There may be experiments unknown to me which prove conclusively that perfumes act upon the olfactory nerve and not upon the trigeminal. Thus the explanation falls short under the Law of Parsimony, because I do not know whether it is adequate or not.

In my own mind, it's fallen short, already. For whatever value the thing has, I give it. I won't go out on a limb with it.

The solution that follows is the one in which I myself believe. It has these advantages: it depends on no one experiment; it requires no change in the usual differentiation of odors.

We start by forgetting all that has been written above about perfumes being irritants. We agree that Science is right: they're true odors. . . . Why hasn't the taboo affected them?

Our answer now is: *because of the innate nature of these scents.*

Zwaardemaker and Linnaeus and Herring and Aronson, to name only a few, have classified odors. No classification has got very far, so we may feel full-free to attempt one of our own. And—we'll let psychology help us in the task.

There are three drives activating men. These are: hunger, sex, and fear. Anger is a complex affair caused by any of the three. An animal without food, or deprived of its mate, becomes angry; or it may use anger as a defense mechanism following fear. The human relations: friendliness and antipathy are founded on the sex and fear drives.

Hunger, fear, and sex! Suppose we divide odors similarly. Hunger: food odors; sex: sex odors; fear: animal odors apart from sexual ones. Food odors vary with the animal studied. We know that the carnivora are interested in the smell of flesh; the herbivora: in the smell of vegetation. But—cats are carnivorous, and yet they like very much the smell of stinking goose-foot (a plant) and go into ecstacies over valerian.

Why?

Paul de Kruif tells how the discoverer of the light cure for tuberculosis got his inspiration as he watched a cat in the sunlight. "Cats," he meditated, "cats don't need doctors. Cats just know!"

They know what the scent of valerian means to them, too! It means: the sex-smell. . . . And that's our pointer!

Why do men and women prefer different odors? Why is it that we males like musk, lavender, cedar, sandal—while women like mint, citronella, rose and violet? Why is it that scientists have noted such an extraordinary increase in odor-consciousness at puberty?

Cats know.

Where do perfumes come from? Musk and civet form the base of most of them—and musk and civet are of animal derivation, and are intimately connected with the sex life of the animal. The female musk deer, the female civet cat know what they're for!

And the sweet flower odors—when do they appear? At full flowering, of course—which is the fertilization stage of the plant.

A French scientific journal mentions a young woman who developed an intense love for the odor of patchouli. She saturated herself and everything about her with it. Loss of appetite, depression, insomnia followed—and in the end, she became a victim of neurasthenia.

Oh, the lovely, lovely ladies—how many of them there must be who can afford to buy the perfume that appeals to them most; use it lavishly; and in the end develop neuroses. Then they have to be psycho-analysed; and so substitute one sex smell for another—and are happy until the psycho-analyst pronounces them cured.

"In all the pharmacopoeia of wizardry," lilts Gordon Teall prettily, "I doubt if any charm in power is greater than sweet fragrancy to stir the mind."

And how! Musk and patchouli, says Havelock Ellis, are genetic excitants. So are vanilla and many others. Hampton puts it more delicately: a sweet smell is one that can stir up the instinct of courtship. Redgrove says the same thing: "The fondness of mankind for sweet odors is in some obscure way bound up with the sexual instinct."

When we consider that the source of every natural sweet smell is in the sex-life of some plant or animal; and remember that most creatures are bi-sexual and through all the higher forms have the same need to make or be made fertile—the obscurity, unless we deliberately shut our eyes, is gone.

If sweet smells were an adequate substitute for sex-life, we could find use for them—before marriage; among persons who by vow or ill-fortune can not marry; or during enforced separations between husbands and wives. There would be fewer divorces, and fewer scandals.

Unfortunately, this is not so. Sweet fragrancies are never odors of satisfaction, but of desire. Their use simply accentuates, and centers the attention on, the desire that is always present. And so—they make the biologic urge more urgent.

Sometimes perfumes are used immoderately. Then something breaks —and we see smell-perverts, like the woman who loved patchouli, and like the lovely, lovely ladies with neuroses.

... And why does a woman use perfume in the first place?

The answer supplies two wandering bits of humor: perfumes were primitively used by women, not with the idea of disguising any possible natural odor, but with the object of heightening and fortifying this natural odor.

Now they use the stuff for the opposite purpose, they think.

But they use the same perfumes!

The hair, beside its value in preserving bodily heat, is supposed to have another function: the collecting of sweat so as to heighten the odor for sexual ends. To this day, hair is most efficient for this purpose.

Women wash their bodies often—but their hair infrequently!

Because this or ten thousand volumes like it will have no effect—and because there may be feminine feelings hurt by what I have written above, I offer to the ladies, by way of apology or something, a bit of information. What follows is of no value to science, though it is absolutely true.

I quote Hazel Rawson Cades: "Every person has an individual body-odor to which perfume reacts, but what this reaction is, takes wearing to determine. The influence of perfumes is not merely imaginary. They are capable of affecting both your mental and physical states. They may depress or exhilarate you, make you feel frivolous or demure, pensive or gay.

"Their effects on other people are not always predictable. They may stimulate, intrigue, annoy, fascinate or make people absolutely sick. They may attract irresistibly or irritate so indefinably that people never realize it is your perfume and not your personality that is estranging them."

The warning is time-hallowed, frequently-voiced, and true. For you ladies, though, it doesn't change the "Go" signal; it merely switches a disturbing red light thereon. Definitely disturbing—for the "wearing-to-determine" formula is like finding you've chosen the wrong chapeau, *after* you've joined the Easter parade.

History offers a helpful hint: Josephine liked musk, which Napoleon damned in good round military phrases. He himself used eau de Cologne in preposterous quantities.

Men and women like different perfumes! Hidden here is the beauty secret, the bit of information promised: since perfumes are sex-odors; and since they react both on the user and on other persons; and, in this latter case: differently on the two sexes—a woman must use extraordinary care (if she *will* use perfumes) in selecting the ones least harmful, and most helpful.

I say: *"ones."* for it is impossible for any single perfume to be correct always. A women, left to herself, will select the one which appeals to her most strongly—and it will be the fragrance most nearly like the body odor of the ideal man, the one she loves or could love.

Because it is a male sex-smell, the perfume will irritate other males —which is why Napoleon damned the musk that Josephine used.*

If a woman gets a man to do the selecting, she will wear an odor that will be a female sex-smell—and though attractive to men, it will alienate other women. More importantly: if it is not the precisely proper accentuant of her own aura, it will be an irritant to her. For unless the man who chooses the fragrance is in love with her—and, in addition, a connoisseur of perfumes—the scent chosen will be simply one nearest the personal body odor of some other woman. . . . the woman who is the selector's ideal.

The obvious solution is: to meet a perfumer and make him fall in love with you before he selects your ambrosial concomitant—and the difficulty here is that the perfumer usually prefers his perfumes.

So all that can be hoped for is a compromise. Have the man who loves you pick out several fragrances which he feels would mingle nicely with your own aura. Out of these, select the one least offensive to yourself, and you'll have, as nearly as possible, the perfume to draw men.

What you must understand clearly is this, though: you need, not *one* individual perfume, but *two!* One to wear at dances, parties, all the other times when a man is more important in the scheme of things. For occasions when other women count most—bridge-luncheons, sorority banquets and so forth—you must have another perfume—your present favorite, probably.

As to mixed parties, the crystal remains cloudy. You might try going as yourself, and see how well you're liked.

"The best perfume," said Thackeray, "is no perfume at all."

We go back to our thesis. Either sweet fragrances, when the taboo was established, managed to avoid it because they at least could be described in any vocabulary—as: the odor of the rose, for instance—or they broke through it, as sex has a way of breaking through any taboo. For at a certain period of life the sex-drive becomes overpoweringly strong, and all taboos and inhibitions go to pieces. And these perfumes were bodiless sex-smells, peculiarly adaptable to sublimation—and any human relation, no matter how gross, that can be sublimated, is in that state freed of taboos.

Sex is taboo. Love is not. Marriage is become a holy institution. And

---

* The historians say Josephine wasn't in love with him.

the sex-odors of animals and flowers are sweet fragrances—of which the poets sing!

Our division of smells was: (1), food odors; (2), sex odors; (3), animal odors, apart from sexual ones.

"Faint odors," Parker says, "are the means whereby animals scent their food, find their mates, and avoid their enemies."

The two classifications are the same, you see. The human animal, even though possessed of all the olfactory acuity I claim, uses his nose for these three purposes and for these three only.

And which does the taboo affect? Food odors, it seems, not at all. They were susceptible of ideation at the dawn of words. Any small shadings of liking and misliking occurred after the food had become the human animal's own personal property—i.e., after it was in his mouth or down his gullet. Objective food odors also escape. Roast aurochs was roast aurochs to every tribesman. For hunger was at work, rhythmically contracting the stomach walls, sending smell messages—not to the brain—to the nose. The urgent odors of the alimentary canal were of paramount importance at the moment—and external food smells fused with these, pleasantly, stimulatingly.

A man wasn't interested in his neighbor's reactions at the moment. He didn't care if his neighbor received precisely the same food smell or not. *He was hungry!*

The taboo left food odors—external and internal—alone.

The sex-odors? Here the taboo gained a partial victory. One man's darling had an aura that, apart from the general female sex-scent, seemed fairly tainted to his best friend. This second tribesman, though, had by now lost all confidence in his olfactory judgment. "Old Oomph knows his wild onions in other ways," the doubt persisted. "Maybe I'm overlooking a bet. Maybe that woman isn't such a spoiled darling, after all." In distrusting his own sure sense, he was like a vacationist who, hungry and lost, finds in the woods a bed of fungi. He's positive they're toadstools, unfit to eat—but they just *might* be mushrooms.

For primitive man the taboo did this: it took all personal agony out of the uncertainty.

—Why should this man have chosen that woman? Why should any man have chosen that woman? ... The mystery continues to this day, but—thanks to the taboo—it's impersonal, and low-comedy, now.

The smell that *was* no mystery, the general sex-smell, underwent a kind of synaesthesia. Another sense got the credit. A woman was a woman because she had a womanly appearance, man said. But if the man was blind, a third sense received undue laurels. A woman was a woman because she *felt* like a woman, the blind man said. And all the time woman moved demurely about, driving men mad with the fragrance of her sex, and being driven the same way by the scent of man.

Only the impersonal, unrecognizable sex-odors escaped. And they had to be sublimated into sweet perfumes for poets to sing about. When the poet turned seer, when a Goethe forgot his taboo and spoke of the human flower, he was praised for his lyric fancy.

And what about animal odors apart from sexual ones? The odors of fear-and-unfear, or liking (or, as some say: recognition-and-strangeness)?

As to this group, the taboo was—and is—absolute. Its real source was in this question of familiarity or strangeness. For here, man discovered the seeming fallibility of his nose; here were the first beginnings of the mystery.

So far, we've discussed these odor-types. (1), food—external and internal; (2), sex—real (individual and general human) and unreal (the musk, civet and other-base flower scents); (3), fear—animal odors apart from sexual ones.

What's left?

None of the synthetic scents have the animal or flower origin, being coal-tar derivatives, mostly. Also, there are certain foul smells, chemical stinks.

As to these: the synthetics have different makeups from the natural ones, but the resultant odors are the same, and it is these odors that the nose reacts to, as sex-smells. The chemical odors, the foul odors are called such only because of the paucity of language. They are irritants, affecting the trigeminal nerve.

There may be, in these foul smelling substances, components which also affect the olfactory nerve—but these are certainly not the powerful stimuli which cause a burning sensation in the membrane of the nose. Even if these lesser components were to be classed as conscious odors, they would have the effect of unconscious ones because of the much greater reaction of the common chemical sense. Similarly, in perfumes, there may be components affecting the trigeminal nerve-ends. Certain two-sense reactions are strictly in order. We put mustard (oil of mustard: an irritant) on bread and meat (food: true-taste odors) every time we enjoy a ham-on-rye.

In our discarded, pro-Magendie argument, it may be, we simply considered the lesser trigeminal part as the whole.

Parker spoke truly of the need for odor and irritant separation. Too long, our case has been analogous to one in which, say, we are ignorant even of the existence of a sense of hearing, and have been trying to explain all the recognizable effects of a moving train, simply by the sense of sight.

Once and for all, we must realize our possession of two distinct senses: a trigeminal (common chemical) touch sense; and a true olfactory sense. Man's misfortune has been that both organs are within the nasal cavity. That is why, for so long, he has been unable to distinguish between the two.

It is with the true odors—the more striking effects of which cannot possibly be attributed to the common chemical sense—that we are concerned. And these comprise the food-smells, certainly, and the sex-odors and the odors of recognition and strangeness.

Whether perfumes stimulate the trigeminal nerve; or as general sex-excitants stimulate the olfactory (or do both): in no case does man's consciousness of them disprove the existence of true or other true odors, subconsciously received.

Final proof of these depends on the light their postulated existence throws on the dark places in man's mind. And so our method, henceforth, will be to assume the truth of our hypothesis, and attempt to brighten certain shadowy spots thereby.

CHAPTER EIGHT

OBJECTIVELY, MAN is a most notable inventor. Subjectively, he's a terrific flop. On his one attempt thereat, he invented a taboo—and the thing became an inner knife to dissever his mind.

I do not think any of the other animals are troubled with such mind-division—unless apes too have language.

An animal's instincts are conscious ones; its conditioned reflexes are always effected consciously. Its mind is of the primitive, unit-type.

Admittedly, from such a type, man's brain came. And in every case where the mind acts primitively—that is: whenever the so-called higher brain centers are not brought into play, man is acting essentially like an animal. And anything we learn from animal behavior will be of value in the study of like behavior in man.

Lloyd Morgan's Canon, which is simply a re-statement of the Law of Parsimony, should be applied with equal emphasis to all animals, including the human ones: "In no case may we interpret an action as the outcome of the exercise of a higher psychical faculty if it can be interpreted as the outcome of the exercise of one which stands lower on the psychological scale." In other words: we may explain man as a god *only when we can not explain him as an animal.*

Certain actions of man are patently so explainable, though curiously enough, these are termed: not animal actions—but vegetative ones. And it is the vegetative apparatus—primitive in form, primitive in its working, primitive in that it requires, so far as is known, no use of the higher brain centers—that we discuss now. Here at least there is no vagueness. Man has it; man uses it. Animals have it; animals use it. Anatomists describe it; physiologists note its effects. It is older than the cerebrum by untold millions of years; older than the spinal cord and backbone; older than the muscles of movement; older perhaps than nerves.

Out of the mother sea the life that is in us came. The ancient salt water of it is in our blood; it *is* our blood—though colored now with haemoglobin. And—since some of the vegetative system's messages are carried along this sea-water roadway—we must discuss it, too.

—And what have the vegetative apparatus and the blood to do with hypotheses of smell?... A great deal! They provide the bedrock of fact beneath the architectural theory; they furnish the solid answers to the questions: When? Where? and How?

After all, the first thing a theory must do is: *work.* The human body can't be altered to order. It's the theorizing about it that must be tailored to fit. Before we can even start to test our hypothesis working, we've got to show *how* it works: the actual, believable way of its performing inside the established structure of the human tenement.

A hundred years before Darwin, a Scottish nobleman evolved the theory of Evolution. He had little proof as to *how* it worked, and the scientists of the period laughed at him. To this good day the books describe him as the "mad Lord Monboddo." Harvey discovered the circulation of the blood, and probably died wishing he hadn't. He didn't know about capillaries; he couldn't explain how the blood passed from arteries to veins. Apparently the stuff materialized out of nothing at veins' ends; made one maiden voyage through the heart-pump; then disappeared into nothingness again where the arteries stopped.

The idea seemed ridiculous—and Harvey, its author, got his full share of ridicule. . . . He confined his research, thereafter, to a noncontroversial study of baby chicks.

For twenty years a Cuban physician, Dr. Carlos Findlay, insisted that the stegomyia mosquito was the carrier of yellow fever. For twenty years his rum-and-rhumba world laughed at him—and continued to die of this same Yellow-Jack.

Yet all the time, Findlay's belief was correct. "The most striking example in history of scientific clairvoyance," Gorgas terms it. For out of the innumerable possible causes of the hot-lands scourge, the bewhiskered old Cuban had picked insects; out of the hosts of the insect kingdom he'd selected mosquitoes; out of the eight hundred kinds of mosquitoes he'd unerringly named the tiny killer: stegomyia.

But—there is a queerly vital time-factor: only by biting a patient during the first three days of illness can the mosquito become infected; twelve days must pass thereafter before the germ becomes virulent in the insect's body. Findlay knew nothing of this; consequently his experiments, over and over again, failed. . . . It must have been divine madness that made him persist in, and keep on expounding his belief, all those weary years.

. . . Until the right meeting of minds. Few wars in history have had such a happy ending as the Spanish-American conflict. For it brought to Havana a Dr. Carter who, in his isolated Mississippi hamlet, had noted the two-weeks' time-lapse between first and subsequent cases of the fever; it brought Walter Reed and Lazear and the other army medicos, desperately ready to test any theory (even the old Cuban's) on their soldier guinea-pigs.

And so Findlay was luckier than most men with an idea. He lived long enough to hear the world stop laughing.

A theory may be simple, straightforward, and true—but until there is blue-printed proof as to *How* it works, it remains inadequate. Necessarily, therefore, we shall dwell at some length on the working machinery implicit in the smell-hypothesis: those ancient knots of nerves, the blood, and two ductless glands.

In man's embryological beginnings there is, immediately after the development of the fertilized ovum into a minute cluster of cells (the morula), a change into the blastoderm, in which the embryo forms a hollow sphere. There is infolding; the embryo consists of two layers; and a further infolding which forms an additional layer. . . . And there are the famous (the books say) and important three germ-layers: the outer or ectoderm; the inner or entoderm; the middle or mesoderm.

Each germ-layer turns into certain parts of the organism. The ectoderm, in which we are especially interested, gives rise to the skin and skin-accessories, the entire nervous system, the special sense-organs, the pineal gland, and parts of the pituitary and adrenal glands.

Why do all the sense organs and the nervous system evolve from this one layer? Dorsey explains it: "As the entire outer surface of the amoeba is sensitive, so man's entire outer layer is potentially nervous." The special sensation organs are on the outside because since the beginning the organism has had need for knowledge of the outside world. Even before nerves were developed, this basic need was provided for, though the entire stimulus-and-response circuit consisted then of an irritant surface sensation causing a muscle to make an avoiding movement.

From the outer germ-layer come part of the pituitary: the posterior-pituitary; part of the adrenals: the medulla; the nervous system; the special sense organs. I ask you to remember that these have a common source.*

And I ask you further to remember that the first of the special sense organs to be developed was the sense of smell. For proof of this, we leave the amoeba, and go forward some millions of years to the flat-worm stage of life. Here the organism has learned to move itself about to an extent. Its front portion thus encounters the outside world first and most often. And so—this front has developed a specially sensitive region reacting to chemical stimuli, the direct ancestor of the modern nose.

In still another way has the worm grown less primitive. It has found that muscles are not enough, and in each of its segments has developed

---

* In adults, the pineal gland becomes useless—and lifeless—brain dust. Its addition to the list above would be of no value.

a plexus of nerves: a ganglion. Two of them, in fact. These ganglia are strung like beads on two strings, up and down its length—and the strings are nerve-connections.

In man, today, we discover precisely the same two strings of beads, outside and in front of the spinal column. So long as Nature finds a use for any created form, she permits it to remain. And the autonomic system (called also the vegetative and the sympathetic system) is useful and remains—primitive as ever—alongside of its neighbor, the spinal cord which is so many millions of years younger.

Let's look at our flatworm again. One nerve plexus is much larger, more highly developed than the others. It is the one in front, for this is the one most in use. So far as the organism has a higher nerve-center, this is it; so far as the creature has cerebral control, it rests in this ganglion. The others are subsidiary to it, ordered by it. Remember: the worm has only one special sense organ. The ganglion is a development of this organ, grown from it—some say: part of it.

A dog's brain, with the cerebrum removed, reveals this same ganglion. Now it is developed marvellously. It has become what is called the nose brain (rhinencephalon*), the region—as Elliot Smith says—pre-eminently olfactory in function.

Take the cerebrum out of a man's skull, and beneath it you find, much smaller—as the cerebrum is so much larger—but still *complete, intact,* the nose brain once more!

Here is a distinction we should emphasize. The other special senses are receptors. They transform stimuli into nerve-impulses, and transmit these impulses to the brain through single nerves.

The sense of smell is different; it is a brain itself. Any division between sense organ and brain behind it must be purely arbitrary. Only because the transformer part of the sense is in the very center of the skull do we term it part of the brain instead of part of the sense organ. The thing can be re-stated in a much more astonishing way: it is simply because the olfactory cells extend into the nose chamber that we call them collectively a sense organ of smell instead of regarding them as an actual portion of the brain.

The other four sense-nerves enter the cerebrum through a common switchboard: the thalamus. The sense of smell has its own private connection. As an analogy, we might think of the cerebrum as a feudal king, one, say, like Louis the Eleventh of France. The visual, aural,

---

* It consists of (here are some long words to forget) the olfactory bulb, its peduncle, the tuberculum of factorium and locus perforatus, the pyriform lobes, the paraterminal body, and the whole hippocampal formation.

tactile, kinaesthetic and taste-senses are absolute vassals. The sense of smell is a semi-independent ruler, like Charles the Bold—nominally a feudatory, but giving little more than lip-service and an acknowledgement of overlordship.

But where is its Burgundy? The cerebrum has its subservient workers: the striated muscles; its means of communication: the spinal cord; its efficient soldiers: arms and legs. What does the nose-brain possess?

Plenty! It has workers of its own—only their muscles are smooth. Its means of communication are two: the blood and the autonomic system; and its twin soldiers can rout the cerebrum's entire battle array, any time.

It's a kingdom within a kingdom.

By our theory: the automatic actions of the body are smell-responses. It is true—and accepted—that the sympathetic system controls these actions. It is not accepted, but it is equally true, that the system itself is under nose-brain control.

Various things tend to establish this. Nothing in man's anatomy prevents. Topping the double bead string are the basal ganglia, large masses of nerve cells beneath the cerebrum. There are definite, easy-to-follow nerve-routes between the olfactory center and these ganglia. And this is natural. Long ago, the brain consisted solely of other ganglia between the two.

In the flatworm there is direct connection: nose-brain to ganglia-string, with no other brain and no other sense organ for control to be attributed to. In the lower animals, in the macrosmatic ones, the nose continues its control over the impulses that pass through these ganglia.

Isn't it more logical, more reasonable to believe that a series of actions originally connected and caused one by the others still follows the same path, than to assert that with the same forces activating, the same results obtained, the connection and causation are lost?

Here we have the most primitive of the special senses—and the more primitive nervous system. We have the special sense least altered and the nerve form least altered—sure sign that both worked satisfactorily. We have a sense organ and a nerve structure both antedating the brain. We have stimulus and response based on the wellspring of life: chemical reactions. And this seems strikingly significant: we have nerves in the two *differing from all other nerves in the body.**

---

* The axones of the sympathetic neurones and of the olfactory have a sheath of Shwann without a medullary sheath.

And finally: in the nose we have the only possible sense-organ that can exercise control over the pathway by which most, if not all of the messages to the sympathetic ganglia are sent—a pathway (once more the primitive!) older than the nerves: the pathway of the blood.

Close-clustered in and about the olfactory regions of the brain are three, and in higher animals, four body-control centers of prime importance.

One such is the respiratory center. The organism breathes fast or slow ... Why? ... Because the respiratory center so orders it. And how does the respiratory center know how fast the animal should breathe?

It's interesting. Either you swallow my explanation—or you gag over the common scientific one, which explains nothing at all, simply changing one mystery into another.

An excess of carbon dioxide in the blood, says Science, causes the organism to breathe faster. Fine! but how does the respiratory center learn that there is this excess? Science doesn't know; supposes the existence of a sense in the respiratory center reacting solely to carbon dioxide. There is no sign of any sense organ; no scientific explanation how the organ works; just recondite razzle-dazzle that explains one mystery into two.

There is no such sense-organ! There isn't even any need of one. Consider it reasonably: messages are sent from the respiratory center. They are either created there; or received there and sent on. Since there has been found no sense-organ—an absolute necessity for message-creation —the logical thing to believe is that the center is simply a co-ordinating and transmitting station for the messages it receives. There is only one source from which it can receive such messages—and that source is the sense of smell!

How? The blood contains carbon dioxide, which filters out through the thin membrane of the lungs as oxygen filters in. In exhalation, any increased carbon dioxide content comes to the nose as a definite danger-stimulus. Immediately the nose-brain sends its warning to the respiratory center—and up goes the rate of breathing.

No mystery, no hypothetical sense-organ—just common, olfactory sense.

Respiration is an indirect control over the blood. There are also means of direct control—the heartbeat, for instance—and there is a vasomotor center for this purpose in the brain, close to the respiratory center.

Here also Science, finding no sense organ, has imagined a totally unnecessary one.

Suppose (we choose a situation that eliminates eye and ear as possible factors), suppose there is no external stimulus whatever; only an internal need for the heart to beat faster. This need must of necessity be shown in an altered chemistry of the blood, and this in turn must equally result in a change in the chemical stimulus reaching the nose from the expired air . . . The result? Heartbeat control—exactly like the nose-brain control of respiration.

There is a third center in the brain of warm-blooded animals: a heat center. The thing's getting monotonous, but for this one, too, Science has improvised a sense organ which is no credit to its creators and no improvement on Nature and on an internally-functioning smell-sense that in this case also seems adequate.

—At normal times, anyway. Fevers are a puzzle. In them the heat centers appear to be thrown entirely out of control. Somewhere in our theory there ought to be an explanation—but I do not know it.

This does no violence to the theory. Neither can Science explain fevers.

What we said about the respiratory center being controlled by the excess carbon dioxide stimulating the sense of smell is so manifest that it is extraordinary Science has never discovered it. That the membrane covering the lungs offers no impediment to the passage of gases is proved every time man draws a breath. That it offers no impediment to emanations other than the familiar air and carbon dioxide mixtures is proved every time a doctor detects the characteristic odor of a respiratory disease.

And—grouping together both respiration and heartbeat: perfume, an external stimulus, admittedly controls both—*through the nose* . . . and not through any imaginary sense-organs!

We turn now to what man's own veins and arteries reveal as to the importance of certain small structures in the body.

Size for size, three organs receive a much more abundant blood supply than seems their just and due proportion. If you visited an army kitchen, and saw beefsteaks and cream pies being sent to three dugouts, while the rest of the future Legionnaires were being doled out canned salmon and hardtack, you'd say the occupants of the three were generals at least, wouldn't you?

The organs we speak of are treated like that. Seventy-five years ago, the physiologists would have told you that not one of the three had anything to do with the body's functioning . . . *Now*—!

One is the pituitary. "And its abundant blood supply," says Berman, "emphasizes its vital importance." Another is the medulla of the adrenal glands, concerning which Berman tells us: "At every age, the amount of blood passing through them is very large compared to the size of the adrenals. Their tremendous importance in the body economy accounts for their being so favored."

The pituitary is really two glands. Of the functions of one of them, the posterior pituitary, Berman reports: "Its secretions control the tone of the tissues, of involuntary or smooth muscle fibre, of the blood vessels and the contractile organs of the body, like the bladder, the intestines, and the uterus. It controls the salt content of the blood upon which the electric conductivity and other properties depend."

Of the work of the adrenal medulla he writes: "It has a specific action on the vascular system, the nervous system, the blood, the alimentary canal, and on sugar mobilization . . . It exerts an influence on all smooth muscle innervated by fibres of the autonomic nervous system. It may be thought of as a 'brain' which takes charge of us when we are confronted by emergencies which mean life or death."

The two glands are important enough, are they not?

Posterior pituitary and adrenal medulla—two of the three organs which seventy-five years ago were considered unimportant. The third, like these two lavishly supplied with blood, is the part of the nose that is the house of smell.

And—Berman has nothing to say about this.

But five years from now—or ten—or seventy-five?

Posterior pituitary, in the very center of the brain; and adrenal medulla, the one ductless gland having actual nerve-connection with the autonomic system—the two are Big Shots now, in all conscience. Kempf and Berman and Cannon and many another will tell you how big.

Because of the sudden paramount importance of these and other endocrine glands, the sympathetic (autonomic) nervous system has of late become important, too.*

Our hypothesis holds that this system is under nose and nose-brain dominance. There is an opposition theory, generally accepted, supposedly scientific, of cerebro-spinal control.

Let's look at it.

The sympathetic nerve-chains (trunks, Kempf calls them) extend from basal ganglia in the brain all the way down. As basal ganglia are connected† to the great gray mantle, one would think that if there actually was cerebrum-control of the ganglia-chains, this connection would be the route such control followed.

It's a perfectly good road; it's the most direct road—but all Science can do is: look at it wistfully, see all the advantages of it, and then pass it up to select, instead, a roundabout spinal cord detour.

You see, the favored sense receptors, eye and ear, send their messages always via this cord. Science, assuming as self-evident the dominance of these senses, must therefore follow wherever eye and ear messages lead. That means: it must map out, for cerebral control of the sympathetic system, a spinal cord pathway.

And again we have an explanation that doesn't explain; mysteries where there is no need for them.

To begin with, Science made a simple thing more complex by dividing the autonomic system into two parts: the sympathetic and the parasympathetic. Even this division is mysterious, for the sympathetic is in the middle, and the parasympathetic (the cranio-sacral system) is on either end. You get the effect of a man who, while admitting to his bedfellow that he is entitled to only one-half the bed, insists on taking that half in the middle.

The division is based on the relative (with respect to the vertebrae) point at which certain small nerves, the pre-ganglions, leave the spinal

---

* Kempf describes and divides it: "(1) A chain of sympathetic ganglia running parallel on each side of the spinal cord. Ganglia appear on each of the trunks at fairly regular intervals. Each trunk runs from the second cervical vertebra to the first piece of coccyx. The two trunks unite in one ganglion at the coccyx. It will be understood that these ganglia, as do all the sympathetic ganglia, lie wholly outside the central nervous system. (2), The cephalic or brain portion of the sympathetic consists of four main ganglia on each side, not appearing in regular segments as do those considered above—the ciliary ganglion (ciliary muscle of the eye, sphincter of the iris, etc.), sphenopalatine (vasomotor, secretory), the optic (vasomotor, etc.) and the submaxillary (glands of mouth, etc.). There are also numerous small ganglia. (3), Ganglia scattered through the visceral organs, in the cavities of the thorax, abdomen, and pelvis, the heart, lungs, liver, alimentary tract, pancreas, and sex organs.... We see the system in operation in the bristling of hairs, dilatation and contraction of pupils, secretion of saliva, inhibition and acceleration of the heart, flushing, gooseflesh, peristalsis, defecation, urination, tumescence of sex organs, etc. The sympathetic neurones, thus, are motor, and control the so-called vegetative functions."

† By two kindred areas, the midbrain and interbrain.

cord and go towards the sympathetic ganglia.

I give you now some of the pseudo-science which the books say is true.

These pre-ganglions, Watson avers, are actually the ones which control the action of the sympathetic ganglia, and they are a part of the cerebro-spinal system.

The sympathetic, he says, has no effective control of its own.*

According to him: the cerebral messages move along spinal cord and pre-ganglions to the ganglia which through their post-ganglionic fiber causes the motor or gandular actions. The sympathetic nerves from the basal and other ganglia go along all the way; pass alongside or through the ganglia and on to the glands—for no reason at all. They don't do anything. They're just wandering nerves with nowhere to go and nothing to do when they get there. And eye and ear are still triumphant—and cerebro-spinal system is still king.

But—why these sympathetic ganglia are so nicely connected by primitive nerves all the way from brain to the coccyx that was once a tail; and why these nerves should wander through the very ganglia you would expect them to wander through if they had messages to carry; and why they should end in the very spots you would expect them to end in if they had any reason for ending anywhere—all these questions are mysteries which Professor Watson hasn't found time to answer.

Nor Berman. His statement is: that in spite of the work of certain investigators (whom he names), recent research tells us that the pre-ganglions control the ganglia.

After you've read quite a few books on popular science, you get to the point where every time you see those two words: "recent research," you feel as if they are red flags and you're a bull. For you know that what you're going to read next is one of those things. I remember, when I first began study in connection with this work, how frequently in one of the first books I picked up, "Psychology from the Standpoint of a Behaviorist," I think it was—how very frequently I encountered those two words, and how believingly I read what so glibly followed about personality or awareness or any of the other things Behaviorism

---

* In Watson's own words: "The peripheral afferent system of the brain and cord (spinal and cranial ganglia) affords the sensory innervation for the tissues controlled by the sympathetic. The peripheral process from a spinal or cerebral ganglion cell, instead of running towards the skin or kinaesthetic structures, turns to enter the white ramus and pass along with the pre-ganglionic fiber. But, instead of ending in a sympathetic ganglion, where the pre-ganglionic fiber ends, it passes through or alongside of the ganglion without making functional connections and ends in the motor or glandular fiber that is under the control of the post-ganglionic or sympathetic fiber."

can't explain. It so happened that the next book I read pointed out these deficiencies, and I lost something, a certain belief in Science with a capital "S", which I fear I shall never regain. But the real blow-off came six months later. In my ignorance, I asked for a book on neurology at the public library, and received one printed nearly a hundred years ago. In it, I read—that *recent research had definitely proved that diabetes was caused by a lesion of the brain.*

Since then, every time I see the phrase, I look for the joker. Let's look for one in this eye-and-ear, pre-ganglionic control.

To begin with—why should there be ganglia at all? If a motor nerve is carrying the message, why shouldn't it, as the undoubted cerebro-spinal nerves do, connect through a synapse—a junction point—with a simple nerve cell which will carry the message direct to gland or organ? ... The ganglia lengthen the period of the nerve impulse, Kempf says. ... The nerve impulse lasts quite long enough to get its order obeyed, everywhere else in the cerebro-spinal system. You get the effect of a lot of unnecessary *working* machinery—and old Mother Nature isn't famous for leaving a lot of *that* lying around. She's careless about removing her junk, true enough, but working machinery generally has work to do.

Next, when you look at the ganglia and then compare them with the pre-ganglion nerves which Science says controls them, you get an impression of tremendous dynamos hooked up by tiny copper wires. Such big machinery—and such a little wire! Those pre-ganglions are small, uncomfortably small to Science. Watson apologizes for them—and when a rough-rider like Watson, who throws out mind entirely because he can't make it fit in with his theory—when *he* apologizes, you know an apology's badly needed. And it must be embarrassing to a neurologist, following motor nerves all over the body and noting how nicely their size is adjusted to the work they do, to find in this one instance, motor-nerves—too small, elsewhere, to make a muscle lift a finger— here charged with the tremendous work of the body's functioning.

But—says Watson—they're motor nerves, and little or big, they must run the motor.

There's another thing: the synapses, the jump-spark connections between those pre-ganglion nerves and the ganglia. Everywhere else in the body they're regular—and any halfway powerful nerve-impulse can jump them. Here, they're of different lengths, some of which it would seem only a champion broad-jumper could manage.

Why?

That's the echo you hear.

So many difficulties and discrepancies! . . . There's no need of them. Long before ear and eye evolved, long before the cerebro-spinal system developed, the body had its ancient way to control these ganglia. It still has it; it still uses it.

Control by means of the blood!

You think this old message route is no longer used? Science admits that the kidneys are directly controlled by chemical substances in the blood (and only indirectly by neural impulses); it admits that glandular action between the various endocrines is accomplished through the self-same means.

Two of these endocrines are our present concern. For they are smell's twin soldiers—the original exponents of chemical warfare. Remember the two ductless glands, the only ones (except for dead pineal) which like the special senses evolve from the embryonic outer germ-layer? They're posterior pituitary and adrenal medulla—two of the three body organs (let's point out again that the nose is the third) most favored by Nature with an abundant blood supply. And of these, posterior pituitary is situated directly behind the nose-brain and possesses a direct nerve-connection with it; and adrenal medulla, as we discovered, is the one ductless gland innervated by the sympathetic system, nose-brain controlled.

And posterior pituitary and adrenal medulla between them can make the entire organism jump through a hoop. When *they* send messages, they send them in a way to be obeyed, for they shoot adrenalin and pituitrin into the blood, and these two are plenty strong medicine. Adrenalin will bring the dead back to life, sometimes, but not when they've been killed by pituitrin. Take a cupful of adrenalin and put it in a pond of water, and you have deadly poison. Take a teaspoonful of pituitrin; put it in another pond of the same size—and you've fire-water beside which the adrenalin mixture will seem but a lukewarm chaser.

These two in the blood—and the organism climbs trees. For the blood carries the news to every part of the body—and every part of the body obeys.

How about the little cerebro-spinal nerves, the pre-ganglions? We can picture these doing the work they're big enough to do, carrying small remonstrating messages from consciousness to viscera that are all hopped up with adrenalin, and doing things of which consciousness doesn't approve. Messages like: "Please don't jump through that hoop!" And: "Please don't climb that *big* tree! It's too high. At least, pick out a smaller one." And finally, on a despairing, plaintive note: "Please! Can't you hear me? This is consciousness speaking!"

It's little attention the body pays!

Posterior pituitary is so important it has the most extraordinarily well-protected position of any organ in the body: a skull of its own—the so-called Turk's saddle—in the middle of the head. And perhaps a release of pituitrin in the blood is all that is needed to set off the adrenal medulla, without any neural impulse at all. If so, posterior pituitary is truly the big boss of glands.

Some of its powers we've noted. Amplifying the list: Science accredits it with a curious control over the sex organs functioning as endocrines; control over respiration and the vasomotor system; control over the unknown periodical phases of the body.

Whatever controls this gland—along with adrenal medulla— controls the automatic actions of the body. As possible proof of our hypothesis, I pointed out posterior pituitary's position in the midst of the olfactory area of the brain; its direct nerve-connection with this area.

But—the optic nerve is close by. And since both pituitary and optic nerve have cerebral connections, a message route between is entirely possible.

And so . . . Smell or sight? Which has the control?

I think we can determine the answer. And to make our conclusions certain, we'll check up on possible control by the sense of hearing at this time, also.

Just what does the eye see? . . . Shape, color, movement—and that is absolutely all. For between the nerve cells of sight and the outside world there is a crystalline lens, through which rays of certain wave lengths only can pass. These are always light rays—and the permitted range does not even include all the wave lengths of light. Infra-red, ultra-violet—they're outside the visible spectrum. Within their limits the twin organs of sight: the eye-sense of color, and the eye-sense of form and movement, are extraordinarily acute—the fovea lutea, the spot of clearest seeing in the eye, being, we are told, a gift animals do not possess. True, we must hang our heads in shame when we compare human eyesight with that of some birds—the eagle, for instance. But still, our eyes within their limits are *good*.

After the light rays penetrate the lens, they cause chemical reaction in the visual purple. It is this reaction which is transmitted to the brain as a neural impulse. But—the visual purple is susceptible to but one

THE SENSE OF SMELL 83

kind of chemical reaction, the kind it receives from light; and the nerves of vision, since they are imbedded in this purple, can receive but one sort of impulse: that caused by this single reaction.

Remember: throughout the body, nerves carry news of chemical change, and nothing else! Free nerve-endings are open to any sort of chemical impulse; but nerves protected from all but a specific form of stimulus can carry impulses only of that.

Olfactory nerves have free nerve-endings; optic nerves do not!

Just what does the ear hear? Sound strikes the stretched skin of the ear drum; reverberates against hammer and anvil bones; winds up finally in the organs of Corti—where at long last there are nerve endings to receive the message and transmit it to the brain. Here again we have a wave-sense, reacting to sounds within a certain length-range. There are slight variances in individual limits. Some persons can hear a bat's squeak. Others cannot; find the tone too high. There is, also, a twilight zone of hearing. An organ can fill a church with silent music, and give us the sensation we get nowhere else but in church. Here, the wave-lengths are border-line close to actual sound.

Hearing, like sight, is strictly limited to vibrations within a definite range. For like the eye, the ear has a filter, pad, cushion—whatever we choose to call it—between the nerve endings and the outside world. And this filter—the ear drum—like the eye's crystalline lens, permits only a certain specific form of neural impulse.

We get back again and again to the fact that the only nerves in the human body, free to receive chemical stimuli without limit, are the nerves of the sense of smell.

Here's a thought-provoking problem.

The posterior pituitary and the adrenal medulla, we are told, keep each other in check. The parasympathetic and sympathetic systems, it is asserted, are also in check. But, in each case, *how do they stay that way?*

Suppose: adrenal medulla pours adrenalin into the blood. Pituitary then must continue balancing the effect with pituitrin. What causes the double action to stop? And when it does stop, supposing the organism in balance again, it's in a different sort of balance from normal times, isn't it? And how does it get back to normalcy?

So long as the secretions are in the blood, the organism is in high gear. What brings it back to idling speed?

We can, by the hardest, explain the build-up—the acceleration—as the work of any of the special senses. But neither sight nor hearing can account for the deceleration. The eye can't turn back and look at what is going on inside the body. The ear can't reverse its megaphone; use its telephone receiver as a mouth-piece.

The need here is for an inward sense, affected by internal stimuli. Without one, the human machine would hold whatever jarring tempo it had attained. Any automatic-action theory, vaguely involving gradual deceleration, a coasting-down slowly from too high a speed, simply substitutes big for little, and makes the entire body a sense organ. Any way we look at it, there's need for an interoceptor.

Here again the exponents of visual control find themselves forced to imagine another internal sense—which, as usual, they can't definitely locate.

Under our hypothesis, there's no need to assume anything. The sense of smell is an admitted interoceptor . . . Too much adrenalin in the blood? The nose detects it; sends word back, and the trouble is remedied.

By several possible means. It is known that glandular messages are carried in the blood-stream. There are four ways through which this stream can be controlled: by the vasomotor system; by the respiration; by the admission of glandular secretions—pituitrin and adrenalin; and by the action of the sympathetic ganglia. *The only known sense which can control any and all of these four is the sense of smell.*

Here's another slant on our eye-versus-nose discussion: How can the sense of sight control the sympathetic ganglia when the sympathetic ganglia appear to control the sense of sight? The ciliary ganglion controls the ciliary muscles of the eye, the sphincter of the iris. To maintain even self-mastery, the eye must control this ganglion.

Theoretically, the control is possible, but consider the route it must follow: the eye gets a too-strong stimulus. It sends its cry for help—in the form of a neural impulse—back to a junction point, the thalamus. The thalamus relays it to the cerebrum. The cerebrum, instead of speeding the required remedying message straight back by motor neurons, similar to those already present, sends it instead in the opposite direction —through the medulla, down the spinal column to the third cervical vertebra, where a pre-ganglion leaves the cord and goes to one of the cranial ganglia. From this ganglion, post-ganglionic fibers take the message back to the region of the eye; a muscle is innervated; and what's needed is done: the pupil of the eye contracts.

It's such a long, long uselessly long road. Once you get hold of a sketch of brain and nervous system, and see the route Science says this particular message follows, you lose all respect for the brain. It's such a completely brainless way to do the business—like turning to the left and going around the world to get across the room, instead of turning to the right and taking six steps.*

Perhaps you are beginning to doubt your eyes. If so, I ask you to consider the matter of Attention for a moment, and to try a small experiment.

Suppose, all at once, you have seen out of the corner of your eye—or heard—something vaguely strange and alarming. You turn to face it instinctively, your eyes alert, your ears listening. . . . Make yourself imagine an occurrence like that in the room behind you! . . *Now!* Go through the motions!

Now try to remember whether you were breathing in or out, just as you finished turning. . . . You don't remember? . . Try it again! You find yourself drawing a sharp, quick, inward breath.

Invariably this occurs, no matter how often you repeat the experiment. If you deliberately make yourself exhale, as you swing round, you find that your attention is upon making yourself breathe so, and not on the stimulus you've turned to investigate. In a real crisis, you couldn't *keep* yourself from drawing that sudden, sharp, quick breath.

You may attribute this to the way your neck muscles work—or in-stinct—just as you prefer. But the fact remains: that any time you turn or lift your head to investigate something potentially alarming, *invol-untarily* you focus your sense of smell on the object, too. And if you check by other tests, you learn that either fixed eyesight or a listening attitude is always accompanied by a dilatation of the nostrils to aid the smelling organ.

You look at the object; you listen to it; and you get conscious reac-tions from both senses. Man (we must repeat this) uses every possible sense to find out everything he can about the thing of interest. But the unconscious reactions—the emotions roused by the object—which never reach the conscious mind until they have occurred, are caused by the sense man is unconsciously using—the sense of smell.

---

* The true explanation is no absurd alternative of nose-control. Here, it seems, is simply an instance of primitive, local stimulus-and-response: a surface irritation (too much light) affecting an adjacent ganglion which causes a muscle (the ciliary) to make the proper correcting, avoiding movement.

In the profound emotions, says Berman, especially fear and rage, the heart beats more strongly; the eye sees more clearly; the ear hears more distinctly, and the breathing is more rapid.

He's just fifty per cent right. The heart beats more strongly and the breathing is more rapid. But see what happens to sight and hearing, the so-called paramount sense organs!

Take fear paralysis, for instance. When a man's afraid, really afraid, the first thing that happens is that he grows almost deaf. You can shout at him, and he doesn't hear you. And his eyes! They get glassy and blank. . . . Adrenalin doesn't yet know that these two senses are supposed to control it.

Take burning rage. When a man's angry, really angry, he can't see very well. There's a rush of blood to the head, and he views things fuzzily through a crimson haze. He is "seeing red,"—and that's no poetic expression! . . . And what happens to his sense of hearing? The heightened blood pressure increases the inter-aural noises. He hears a buzzing in his ears, an inner static affecting the clear reception of outside sounds.

It's rough going for the "favored" senses during emotional storms!

In a structure—or a text-book dogma—basically unsound, deep cracks inevitably appear. The builder conceals *his* faults behind a covering coat of plaster. The unhappy dogmatist, in similar case, can only use his imagination—and an air of authority.

To account for various automatic functions of the body, we are told that we must assume the existence of the following sense organs, not one of which has ever actually been discovered:

I. Receptors of the digestive system (exclusive of smell and taste):
  A: Organs or sensory cells of hunger—in the stomach.
  B: Organs or sensory cells of thirst—in the mucous membrane.
  C: Organs or sensory cells of nausea—in the stomach.
II. Receptors of the circulatory system.
III. Receptors of the respiratory system.
IV. Receptors of the reproductive system.

There is the list. And—it would be a big help if the anatomists could just find just *one* of them!

We have already discussed the imaginary receptors of the respiratory and circulatory systems, and found no need of them. As to the reproductive system: it is certain that there is a curious affinity between this system and the sense of smell, and equally certain that the sense exercises some sort of control over the system. There is tissue in the olfactory cleft both in form and functioning exactly like characteristic tissues of the reproductive organs. . . . And there is other evidence, which we omit, to avoid having this book banned in Boston.*

There remain the three apocryphal organs of the digestive system—and these can be disposed of, together. Smell and the relatively minor senses of taste are admittedly receptors for this system. Proof that they are the only ones rests upon their adequacy to serve the system's every sensory need.

Now, each such need has its physical manifestations. And these manifestations (rhythmic contractions of hunger; excess stomachic acids, et al., of nausea; pharynx dryness of thirst) must result in component-changes in the internal odor; must thus become definite olfactory stimuli.

Adequate ones!

Bulwer-Lytton predicted that some day men would live on the odors of foods and not on the foods themselves. It's a fascinating thought. Coffee and tea are stimulants, supposedly because of the caffeine and thein (chemically the same) that they contain. Yet—caffeine is an odorless drug, and inhaling it in its pure form gives no Mocha-and-Java stimulus. We get a pleasant lift from coffee, simply by smelling it being made—and there's a one-in-a-million chance that there may be some sustaining principle in the stuff which if isolated would maintain life in some such way as Bulwer-Lytton forecast.

That's fantasy. This is down to earth:

You come home in the evening, tired and hungry. Dinner's ready, and you sit down and eat it. Immediately afterwards, before any but the very first processes of digestion have started; before the food has had time to do you any good—as food—you get up, not tired any longer, ready for an evening's play. Yet—all you've done, actually, is to smell food and swallow it.

They're not enough to explain the puzzle. Neither is the usual solution offered: that the stomach's rhythmic contractions† have ceased.

---

* See Havelock Ellis: Volume IV, *Psychology of Sex.*
† Cannon's experiments on dogs—to be described later—deal this theory a desirable death-blow.

And if the stomach walls were *all* sensory spots, they'd be no service here.

The real answer is: *that the nose anticipates the body's blessings.* Long before there is any real assimilation, there must be a fusion of internal emanations. With this result: odors of the body's need have combined with satisfying food-scents. And—since the nose deals strictly in emanations, the total inner sensory effect must be a stimulating one of all desires fulfilled. . . . You're ready for an evening's play.

The median processes of digestion are blood-controlled— and we've discussed control of the blood-stream.

The final process—evacuation—is interesting as an olfactory response, too. Maybe I ought to omit this, but—

The curse of civilization, patent medicine fakers and medical doctors agree, is constipation. It is responsible for a whole train of ills, and it is on the increase . . . Why? . . . Because: civilized man has toilet facilities and running water. Under modern conditions of plumbing, there are no adequate smell stimuli and no strong smell responses. I'm trying to be euphemious . . . but my dog gets a definite stimulus from evacuated matter—and responds. The clean scent of soap in the modern bathroom can never be any more than a substituted stimulus. And substitute stimuli, Watson tells us, are never as strong as instinctive ones.

Incidentally, this particular function remains almost the only human-animal action for which some wishful thinker hasn't invented a higher-than-animal motive.

Haltingly but fully, I think, we've covered the ground. But before we end this longest, hardest chapter of the book, let's give a proper emphasis to a point we made too casually.

Dorsey itemizes the obvious effects of profound visceral changes, such as occur in fear, horror, pain, etc. Underlined, phrase by phrase below, is his list. Notice the results these effects produce upon specific senses:

Tongue cleaving to roof of mouth! This means: the taste-buds are put out of commission.

Pupils of eye dilated! No consideration is shown here for the conditions of clearest vision. The pupils, before the dilatation, were adjusted to the brightness of the environment; now they are not! And so too much light is suddenly thrown on the retina, and vision is blurred.

... Imagine going from a dark theatre into a sunny street. The effect is precisely the same.

Muscles of face—and especially of lips—trembling and twitching! In other words, the sense of touch is momentarily out of control. So too, are its auxiliary senses. A man in the condition Dorsey describes has no consciousness of what he contacts; or of heat or cold or minor pain.

Pounding heart! This works particularly against the sense of hearing. As Dorsey himself states: disturbed blood pressure inside the ear gives rise to such noises as ringing, roaring, rushing, and so forth.

Cold sweat; hair on end! Because, perhaps, I am still somewhat bound by the ancestral taboo, still hear ancestral voices whispering: "mustn't say the naughty word!," I have failed to point out that body odor may be withheld—as in the case of nesting birds—or heightened, at need; and a sudden increase in perspiration is in essence a heightened-odor response to a message, blood-brought, from the nose. Such motor-response appears in its highest—or some other adjectival—form in the mephitic skunk; but even in its less shocking, more common guise it remains the organism's warning "Keep Off!" sign to the outside world.

The "hair rising on end" may, it is true, be caused by the organism's desire to appear as large and ferocious as possible (the usual explanation)—but it must be remembered that an animal with fur erect is a more odorous animal, if only because of the larger surface from which the air receives emanations; and that a ferocious odor is probably a better safeguard against a possible animal enemy than an increased size. Most dogs learn to leave skunks alone. Hudson tells us the puma never attacks man because it knows the human odor and fears it.

Hurried breathing! An aid to odor-perception, since it brings more frequent bulletins from the world outside.

These are the observed effects. They reveal, as to the theory of eye-and-ear control, its latent, low-comedy corollary: twin leaders, carefully rendering themselves hors de combat before each major battle.

In emotional crises, one special sense is aided; the others are hindered, thwarted, marred. Remember: we spoke of old servants in an old house, an old queen dethroned, and new mistresses honored. It seems that every time danger looms, there's a palace revolution! The sense of smell becomes a super-sense—because the ancient animal body trusts only its ancient, animal control.

To sum up: stripped to simplicity, the smell-hypothesis holds that what was, *is*—there being no proof of necessary intermediate change. The accepted doctrine seems: that what was, is not; what is, was not— and who said intermediate change was necessary?

Between these two theories, by rule and law, we must "select the one that makes the fewest necessary assumptions to cover the facts."

Our hypothesis states simply that the sense of smell, through its control of the sympathetic nervous system, the blood stream, and two ductless glands, controls the automatic actions of the body. If we reject this; if we try to trust instead the established dogma of eye-and-ear, cerebro-spinal control, we must assume, as a condition precedent, a cataclysm, systemic, organic, sensorial—an inner earthquake of incredible proportions. . . . which left no visible scars. We must use imaginary exits to escape from very real blind alleys; we must accept a dozen internal senses, not one of which has ever been discovered, to supplement a smell and taste which make the others quite unnecessary. We must be deaf to disharmonies; be blind to incongruities; be dumb—and be scientific. We must explain away curious exceptions in the nervous system; we must gloss over the brain's seeming brainlessness in such matters as controlling the pupils of the eye. We must, invariably, make man a god instead of an animal—even in his most animal-like actions. And then—we are stopped short by the ultimate absurdity of a supposedly dominant sense *making itself less sensitive in its most need!*

Not one of these assumptions is entailed in smell control.

By the Law of Parsimony—which theory is correct?

CHAPTER NINE

SOMETIME AGO we charted certain mysteries: instincts, emotions, etc., to explore. Over these the stark civil wars of Psychology have been fought—with no words or experiments barred. Yet they remain enigmas.

As such, they offer now the ultimate test of the smell-hypothesis. For our formula must, if true, furnish for each of them a swift-and-simple explanation.

Because there is a somewhat illogical but nonetheless definite danger in unexpected simplicity, we begin our task with a pertinent bit of imagery.

Suppose:

You've been sent to open a door, only to discover you can't, because it's locked. As you stare at it, a man comes up.

"Here's the key," he says. "I just found it."

Naturally you don't argue over what seems a small stroke of luck. You take the proffered key; insert it in place. It goes in smoothly; seems to engage the inner mechanism fairly; turns—but not quite all the necessary way. You add a little brute force; and then, a little more—but still the key won't complete the desired arc.

Your helpful friend hovers at your shoulder. "Try it the other way," he suggests. "Maybe there are right and left-hand locks."

Obediently you try turning the thing the other way. No luck!

"That's funny," says the kibitzer. "Let me try it!"

So you surrender your place to him. He takes his solemn time; rolls back his sleeves; rubs his hands; at long last bends over the lock.

Now *you're* kibitzing. He jiggles the key back and forth, as far as it will go in either direction; tries to put it in upside down; finds he can't; goes through his former routine.

Faintly the suspicion rises within you: "Maybe it's the wrong key." Eventually you suggest this out loud.

The other man rejects the thought with scorn. He's the key's discoverer; he has a personal interest in the thing by now. In a way, he's its sponsor. And so he points out all the various mischances that might hinder the key's proper functioning: rust or dust in the mechanism; a scarred place on the catch inside; the dragging weight of a door off-balance, and so on. And he has various remedies: tapping the key; lifting the door by its handle while he turns the gadget; having you do this, while he raises the door . . . the thing drags on and on.

And your first faint suspicion becomes a dead certainty. Very firmly you tell him: "It's the wrong key!"—and start doing what you should have done long ago: look for the right one.

In its usual hiding-place, the mail-box, you find it. It opens the door easily, effortlessly, instantly.

You've been maybe twenty minutes doing the job. But—and here's the point of the fable—the actual opening of the door didn't take that long. You simply wasted nineteen and three quarters minutes— *fumbling with the wrong key.*

There are unopened, mysterious doors in the mind. And newly-found keys of theory have been proffered many times to explain them. And every resultant failure has been obscured by the disappointed sponsor's wordy fumbling, long dragged-out, to the effect that under different planetary conditions, or something, his key might work, even though in the immediate instance it manifestly doesn't.

A human habit is: to make a joined pattern of things often found together—and there is a very real possibility that some of you, having grown accustomed to abstruse wordiness always, may think it a necessary concomitant to any attempted solution of age-old mystery; may feel that any easy, simple explanation is plain presto-chango stage-magic, and so unworthy of belief. To such, I would point out that the right key *should* open a door easily, and that this ease of opening in itself should form a proof of the key's inevitable rightness.

Take for example:

INSTINCTS

The implicit hereditary or instinctive responses, the books say, appear in changes in respiration, circulation, and the whole system of hormone secretions.

These changes and secretions are smell-responses! Instincts are no more mysterious than other chemical reactions. Certain chemicals like each other; certain others do not—and the test tube that determines this is: subconscious smell.

The kid that Galen brought into the world before its time was born with the ability to move—and a chemical laboratory of a body. There were compounds in this body that were unstable, restless—*hungry,* if you wish. The emanations from this instability were brought by the blood to the lungs. Here they passed through a membrane into the expired air to act on the free nerve endings of the sense of smell. As sensations, they kept coming to the brain—and the brain sent messages to the animal's body to: "Do something about this trouble! *Move!*" The kid moved—legs, neck, head. Along the row of pans it went, taking a test-sample from—*smelling*—each one. Finally, it found an odor which blended pleasantly with that from the unstable compound in its body. And so—the kid drank the milk.

That's instinct—in animals.

Food instinct!

Instinctive behavior, we are told, is unlearned behavior. It functions with the first adequate stimulus; it is common to man and many higher animals; it is complex; it is accompanied by, but not dependent on, consciousness. It is explicit or implicit.

The first adequate stimulus!

The first the newborn baby gets is the doctor's slap, to make it breathe. Yet it is impossible that this small spanking actually forces air into the unused lungs. Instead, this occurs: the small current of air reaches the free-nerve endings of the nose; becomes their first external stimulus! The breathing that ensues is a motor response.

Perrin and Klein say that the infant does not possess an instinct that drives him to seek his mother's breast. He is put there and conditions are so arranged for him that suckling ensues. They say the proof of this lies in the fact that the infant will respond to a finger or a rattle.

"So far as the human being is concerned," they add, "a careful study of the genesis of feeding behavior seems to show that what is inherited and therefore instinctive is limited strictly to the series of metabolic or physiologic reflexes (the stimulus of rhythmic stomach contractions, tactual stimulus to lips, swallowing) already mentioned. In addition, the inherited nervous connections make for a display of random movements of a restless sort, plus vocalizing responses whenever the internal stimulus asserts itself."

Those movements may *appear* random; they're not! They're purposeful—but impotent. The baby makes his small attempt, but lacks the sheer strength to go find his food—as Galen's young goat did.

The response to a finger or a rattle (tactile stimulus to lips, swallowing) seems mechanical rather than truly instinctive. Put your finger against the roof of your mouth, far back—and nothing happens. Press hard—and involuntarily you swallow. Thus, the swallowing is the result of pressure, and nothing else. Eventually, the baby's response to a finger or a rattle becomes a learned response, a conditioned reflex. But it still has nothing to do with instinct.

The "stimulus of rhythmic stomach contractions" we've considered already. The remaining point: "that the infant does not possess an instinct that drives him to seek his mother's breast," may in a narrow sense be true. What he has is a chemical need for the food that is there. But it is certainly instinct—the smell of milk—that makes him keep on suckling.

For the thing's cross-sectional value, I outline now every experiment on instinct described in one college textbook*. Let's see how our hypothesis affects, and is affected by, each one.

In a discussion of whether emotional stimuli can be inherited instincts, mention is made of James' statement that the fright of a puppy thrown on a tiger-skin rug can be explained in emotional experiences of the puppy's ancestors. We are told, though, that modern biological concepts of heredity hold that the inheritance of vestiges of ancestral experience is now a practically discarded theory. Two experiments by Stratton on cattle are given as "carefully controlled experimental investigations."

He tested the common belief that a red object angers cattle. He found, in both wild and tame cattle, that it did not. He also learned that his experimental findings were corroborated on the whole by the judgment of California cattlemen.

On this, our comment is: Here, the sense of smell does not enter. The experiment disproves a commonly accepted belief already disproved by practical experience; and proves—if a positive conclusion need be drawn—that sight is a poor emotional stimulus to cattle.

As we believe.

Stratton then tested the tradition that blood has a direct exciting influence upon cattle. He states that it does not. He found, in this case, that the cattlemen believe that it does.

---

* The one we've been using: Perrin and Klein's *Psychology*.

I offer this as a proof of how carelessly uncontrolled a "carefully controlled" experiment can be. It would seem obvious to anybody that the blood smell encountered by cattle under natural conditions would be that of horses and cattle killed by carnivora, horses and cattle which, in all likelihood, had previously been maddened by fear and excitement. This blood would have a changed chemical structure, if only because of the tremendously heightened adrenalin and glycogen content, in sharp contrast with the fairly normal odor of blood from cattle slaughtered as quickly and painlessly as possible by men.

The direct, exciting influence to living cattle would be the direct, exciting fear odors in the blood of specific animals chased and killed by wild beasts. The normal blood-scent would be a familiar one, such as the tested cattle received themselves, continuously, from their own lungs.

The cattlemen, who believed they had seen actual proof of blood being an excitant, had seen such proof, because they had seen cattle faced with blood of the excitant sort.

Stratton had not—and his experiment is worthless for this reason.

The "practically discarded theory of the inheritance of acquired characteristics!" Perhaps we should speak of that here. Not that I am so sure emotional stimuli are inherited—but because I did state that a taboo against the sense of smell *is*.

It seems to me that evolution can be simply explained by a chemical theory, and that an actual instance of an acquired characteristic being inherited can be shown.*

In the beginning there was life—a chemical laboratory in a chemical environment. From this environment, life took the chemicals it needed to sustain itself. But these chemicals were never pure. They were in combination with foreign, unneeded elements, always. These, and the end-products of the organism's anabolism-and-katabolism became waste, useless chemicals to be discarded.

Ingestion, metabolism, excretion. This was the life process: chemicals received, transformed, disposed of.

So long as the environment stayed the same, the substances received did not vary, and the organism remained as it was. But—at any change

* I speak here of a *complete* chemical sense: a combination of touch, taste-buds, common chemical sense of the mucous membrane, and the olfactory nerves. Unquestionably all these developed from an original primitive sense of chemical irritability—a sense which, in more-or-less divided form, all organisms possess today.

in chemical environment, new and strange substances entered in. These, in the form received, were irritants to the system.

But in the body there was waste matter, formerly useless; now suddenly valuable. For in this waste the organism discovered chemicals which would combine with the foreign elements, ending the irritation. And so these combinations were made whenever needed. Accidental trial-and-error developed a use for them—and they took on functional permanence.

With every change in chemical environment, a like process ensued, and the outcome was always the same. So long as the organism was stationary, it was completely at the mercy of its environment. As it developed means of locomotion, it could control external conditions to an extent by moving to avoid new irritants. The more mobile it became; later, the better it could swim or fly or walk—the more of its chemical environment it could control, and the less need there was for change in bodily chemistry.

Finally, the highest organism learned to talk—and this revealed a new possibility: communal control. For judgments could be formed and transmitted; the experiences of many men could be collated; any change in environment could, in enormously greater degree, be controlled.

As an example, though not a very good one: A single returned nomad hunter, at the advent of the Ice Age, could, by saying: "It's warmer, south!" save an entire tribe.

And so—today we see man triumphantly static in his air-conditioned, winter-warmed, summer-cooled, sterilized civilization.

The important thing—because it makes any proof of heredity difficult—is this: Evolution is a clock almost run down. It was swift at the beginning. It moves almost imperceptibly now.

Barring cosmic change, we are as we will be.

But the organism has never lost its power of chemical adjustment. Foreign bodies can still be counteracted by new kinds of antibodies. The drug fiend's body is a perfect example of the workings of evolution.

—And because the counter-irritant the body creates to adapt itself to the drug is in the blood, and carried wherever the blood goes, we find babies born drug-fiends who have never seen a drug. The blood in their bodies—which is the same as the blood in the drug-using mother's body —has, like the mother's blood, acquired a new function: the creation of a specific poison antibody.

A (chemically) acquired characteristic has been inherited.

Let's get back to the experiments in our case-book:

Watson made tests on infants. He discovered three ancestral emotions; pain, fear, and love.

And he forgot, all the way through, that the odor of the experimenter should have been taken into consideration in every experiment. For, long before his eyes are opened, the infant uses his sense of smell. As we pointed out, the first breath of life is a smell response. Even before it is born, the embryo may use this sense. Zwaardemaker thinks he proved that odors are diffused into the olfactory clefts, without the necessity for breathing, and that any odor which comes to the nerve-endings there causes a reaction. If so, the unborn baby receives external sense-messages from its mother's body.

—The statement by Perrin that many so-called instincts, such as the employment of language and modesty, are social habits, seems entirely in line with our formula—for the nose has no part in them.

Bernard, considering the family relationships—comprising the maternal, paternal, and parental instincts—reaches the conclusion that they contain two groups of factors: the physiological and the cultural.

And this suits us very well. For our hypothesis can completely ignore the cultural group; and any action explained in terms of physiology (visceral and glandular actions, with certain random outward movements) can be more adequately explained with the sense of smell as sole receptor, the visceral and glandular movements as primal motor effectors, and the random outward movements as secondary effectors.

—Yerkes and Bloomfield made experiments to determine if kittens instinctively kill rats. The two investigators answered this question in the affirmative, but other psychologists aver that the experimental evidence "does not prove the existence of an inborn, integrated tendency to kill mice."

It seems that this is a battle we can happily leave to Yerkes and Bloomfield and the other psychologists. Neither side does any damage to our own theory; neither offers any conclusive evidence of anything.

One or two observations, however, seem worth while. The kittens, though physiologically immature, were able before their eyes were open to react to certain odors: weak ammonia, sour yeast, leather, and the experimenter's hand. These four only are named.

Ammonia is an odor of evacuation; sour yeast, of digestion. Leather and the experimenter's hand were skin odors of strange and, to the kittens, dangerous organisms.

(Freud, who stresses the sex life of infants, may possibly be able to explain why a sexually immature animal is conscious only of odors pertaining to the other two drives of life: food and fear.)

—After the kittens' eyes were open, they paid little attention to the mice until the latter moved. Then their reactions were compounded of fear and rage.

Movement, of course, increases scent. We spoke of certain nesting birds being able to withhold their odor while brooding. It is entirely possible that rodents have some such power so long as they remain motionless. Whether this be true or not, the mouse odor was not, to the kittens, exclusively a food odor. It was also a strange one—and therefore, fear-causing. And it was this fear-odor which evoked the kittens' reactions.

"During its second contact with a mouse one of the kittens caught the animal and devoured it." Even this did not begin as a food-seeking action. It was fear-caused rage—defense by attacking—and the first food sensation ensued only after the mouse was wounded, when the smell of blood came to the kitten's nostrils—to awaken ancestral memory.

Instincts, we may say, are the reactions of a basically adapted organism to each first and later manifestation of the various chemical stimuli to which its body is adapted.

And—they're odor-responses, every time!

CHAPTER TEN

FROM THE INHERITED EMOTIONS, we turn to the self-made ones that color our daily lives. The James-Lange theory explains these as conscious states directly caused by physiological activities. To quote our text-book: "More specifically, the James-Lange theory regards emotions as fusions of body-sensations that are aroused, like other groups of sensations, by the stimulation of sense-organs attached to sensory nerves."

Our own formula of olfactory control is but a revealing, sharply-defined re-statement of this. It is a curious fact that we accept the basic James-Lange theory while denying certain amplifications of it, and doubting certain supposed proofs.

James and Lange identified the emotions chiefly with interoceptive sensations. *We* identify them as internal reactions to smell-stimuli—and specify the interoceptor. But—with their statement as to the steps involved in the *creation* of an emotion, we are in full accord.

They are: (1) The stimulus or exciting situation. (2) The physiological changes And (3) The consciousness of these changes, or: the emotion itself.

I quote: "In the case of an individual terrified by the sudden discovery of a rattlesnake at his feet, these steps are represented by his perception of the reptile, his physiological expressions of fear, and his state of consciousness resulting directly from his physical commotion."

Dorsey describes the same situation as occurring to him, and attributes all credit to his sense of hearing. At the moment of the sound, which consisted of a friend shouting: "A snake!" Dorsey was fast asleep. How consciousness remained with him sleeping, as apparently it did, for his friend did not repeat the words and so Dorsey, asleep, must have had consciousness of their meaning—nobody knows but Dorsey.

There was, once, a false alarm of fire at a house where I was sleeping. Someone came into the room and shouted: "Fire!" I woke up; said "What?" blankly. The second or third time the alarm was repeated, I understood. And this, to me, seems the way all minds work. Words are sounds—with meanings, but the meanings must reach a conscious mind to be understood. Dorsey says he was able to understand the meaning of words while he was asleep. He also adds that it was a cultural reaction; that had his interest been ophidia, and not ethnology, his behavior would have been different. Judgment and a realization of professional interests—all between sound sleep and an unconscious jump! What a man!

Remember Beebe's account of the danger smell of snakes? The shout by Dorsey's friend was a signal for attention. Attention's concomitant, as always, was a quick, indrawn breath. By this breath, Dorsey's nose received the smell of snake. *Then* he jumped—and his jump was the act of a body a millions years older than culture.

The James-Lange theory, so Perrin tells us, has been criticized from its first appearance. These criticisms are of three sorts, based respectively upon (1) logical grounds; (2) introspective considerations; and (3) direct experimental evidence. Let's look into these objections.

(1) The logical criticisms: It is urged that an emotion consists of factors other than the kinaesthetic and organic sensations stressed by James and Lange. In other words, the theory is correct as far as it goes, but it falls short. Hunter finds in every emotion not only its core of bodily sensations, but also an affective element of pleasantness or unpleasantness, and an awareness of some thought or object as the arouser of the emotion.

It is the coming-upon of corroboration like this that continually animates me against the ancient smell-taboo I myself must contend with. Here is a theory of tremendous importance—the only one psychology admits as anywhere near adequate—and *it is attacked logically because it leaves out the very factors our own hypothesis supplies!*

Completely! For the "thought" of which Hunter speaks must itself be a product of some subconscious smell-stimulus.

(2) The introspective criticisms. I quote: "Many psychologists, including those who are generally sympathetic with the James-Lange explanation, find that the theory fails to indicate the differentiating factor in the specific emotions. They note that many diverse emotions are manifested by the same bodily disturbances—and conversely, that the same emotion, subjectively analyzed, shows a variety of physiological expressions . . . That the James-Lange theory is entirely inadequate with respect to this issue can scarcely be questioned, but it does not follow that the discovery of the differentiating factor or factors will militate against the theory . . . (for) granting the existence of specific differentiating factors in emotion, it may still be true that the factor common to all emotions is the background of kinaesthetic and particularly organic or visceral sensations."

It is no more surprising that there should be a factor common to all emotions than that all emotions should be felt in a common body. The systemic economy is not complex enough to furnish entirely different visceral changes for each emotion, particularly since the profound emotions each require visceral changes throughout.

The differentiating factor (repetition seems necessary, here)—is the specific causative smell-stimulus. The reason why the same emotion, subjectively analyzed, shows a variety of physiological expressions, is that the stimulus is always a combined stimulus of objective and subjective odors. As the objective smell-stimulus is altered by the personal odor, so does the stimulus affect different persons differently; as the personal odor is altered by its possible variants, so is the combined stimulus altered. Thus the same stimulus affects the same person differently at different times.

Adrenalin and pituitrin released into the blood cause the physiological and visceral changes. But adrenalin and pituitrin are released on specific orders, they being primary smell-responses like the visceral and physiological changes. The consciousness of these changes, and of their extent, has to do with the profoundness of the emotion felt. The affective element of pleasantness or unpleasantness is a direct smell-message to the hemispheres of the brain.

Another criticism of the James-Lange theory is: that an emotion is really intensified rather than weakened by a suppression of its physioogical manifestations; that silent, controlled grief is much more poignant than this same emotion violently expressed. (The control can be, of course, only an inhibition of the outward, visible expressions, since what occurs inside the body—the smooth-muscle tonicity and endocrinal activity—cannot be controlled by consciousness).

A method of quelling an emotional condition, we are informed, is to neutralize it by engaging in some physical activity. Whistling to keep up courage is a familiar instance.

Our comment: This control can take place only after conscious realization of the emotion. The autonomic system and the blood-stream have already carried their messages. Smooth muscles and viscera are already functioning. Then consciousness interferes—through its cerebro-spinal, striped-muscle system. The result is conflict—and this in itself seems enough to explain the greater poignancy of the controlled emotion. The cerebro-spinal effectors attempt to enforce the conscious interdiction; the sympathetic ones try to continue the work they've already begun—and, in the many parts of the outer body under double control, there is tension, a feeling of acute discomfort—such as a handcuffed man feels when a mosquito alights on his nose.

There have already been released into the bloodstream a definite amount of adrenalin, and a definitely increased quantity of blood-sugar. Certain organs are prevented, by this conscious, outward, emotion-control, from utilizing their due portion of these. The legs, for instance, are not permitted to use up their share, running. Consequently, the viscera and glands not thus prevented—the internal organs that can not be so inhibited—have a greater supply . . . i.e., a supply which will be taken up from the blood-stream less swiftly, and so will cause the emotional sensation to last longer.

No wonder the emotion, thus suppressed, is more poignant.

The example adduced: whistling, is of interest. One result attained thereby is a lessening of the sharp attention that is the means by which the nose receives its clearest objective stimuli. Another point: one whistles through the mouth, not the nose. Because this causes a diminished air-current, expired and inspired, along the usual direct route, fewer and weaker smell-messages are received.

(3) As direct experimental evidence against the James-Lange theory, we are offered the critical findings of the physiologists Sherrington and Cannon.

To quote the textbook again: "(a) Sherrington transected the spinal cords of some dogs in the lower cervical region in the attempt to eliminate visceral sensations from the emotional responses of the animals. He found, however, no alteration in the subsequent emotional behavior, and concluded that organic sensations (or processes) are not necessary components of emotion. In another experiment, using two of the same dogs, he cut both the vagus nerves, thus eliminating possible sensations from the stomach, lungs, and heart. Again the dogs growled when teased and displayed as many apparent signs of emotion as their condition permitted. And to prove that his animals did not react with emotional habits acquired through experience and 'stored' in the brain, where sensori-motor connections were still intact after the operation, Sherrington repeated his experiment upon puppies nine weeks old and obtained similiar results."

Perrin and Klein comment thus: "But the majority of psychologists regard Sherrington's results as inconclusive. He merely established the fact that certain elements in the total emotional response, namely, those unaffected by the operation, can be elicited after the nerve connections between the viscera and brain have been severed. In other words, he did not prove that the dogs really experienced emotion."

To me this last sentence is another one of those things. If animals do not experience emotions, then the hundreds of books on animal behavior and the data of thousands of experiments made on animal bodies must be thrown away as useless. Watson's maze experiments, Pavlov's conditioned salivary reflex tests, Kohler's work on apes—not one becomes worth the paper used to describe it.

The statement is simply not so! For if ever humans act like animals, and if ever animal behavior is of worth in the study of human behavior, it is here.

Whether dogs really experience emotions—and there is a library to prove that they do—or whether Perrin and Klein, and the "majority of psychologists" they mention, wish to re-define emotion as a term restricted solely to humans, but meaning: acting like an animal—all this is merely a matter of word-quibbling.

We go on. Sherrington's experiments are regarded as inconclusive because he "merely established the fact that elements of emotional response unaffected by the operation can be elicited after the nerve connections between the viscera and brain have been severed."

This one fact torpedoes, makes scrap-metal out of any James-Lange theory that accepts cerebro-spinal control.

But one that doesn't, remains intact. Sherrington's research was better than his reasoning. He went wrong in his conclusion: that organic sensations (or processes) are not necessary components of emotion. What he actually established was: that the spinal cord is not the means by which emotional sensations are carried to the brain; that the vagus nerves are not the means; and finally, that emotions are not necessarily "stored" habit-reactions.

The smell-hypothesis denies these three things, also; yet it keeps organic processes as necessary components of emotion. Here is its path of sensation: Olfactory stimulus; messages sent through the interworking sympathetic and blood systems; control center responses and glandular secretions resulting in motor and chemical reactions throughout the entire body. Bloodstream and expired air carry the composite odor that reports these activities back to the nose. As an untabooed, internal-situation message it is received in the hemispheres of the brain, and a surprised consciousness finds itself possessed of a core of bodily sensations, an affective element of pleasantness or unpleasantness, and an awareness of some cause therefor.... In other words: an *emotion!*

By breath and blood and sympathetic ganglia and glands—beginning and ending at the sense of smell—the emotional circuit runs. And—Sherrington's experiments help prove it!

(b) A second physiologist who attacks the James-Lange theory is Cannon. Most of his tests were made on hunger. He finds that the conscious pangs of hunger are accompanied by contractions of the stomach walls, and says that all available evidence points to the contractions as the cause of hunger—or as the immediate condition for the arousal of hunger sensations.

He found, he states, that the sight or smell of food causes these contractions to cease; causes the secretion of gastric juices in the stomach; and causes the mouth to water. He found that the actual presence of food in the stomach is not the cause of this.

The question whether smell or sight is the prompter here evokes—and rightly—our usual answer. With this addition: the experiments were performed on dogs. Smell is admittedly these animals' first and preferred receptor. And it was of necessity a component factor throughout—while sight was not. The fact that the actual presence of food in the stomach was ineffective ought to indicate anew that there is no internal organ of hunger there.

Cannon also demonstrated the inhibitory effect of unpleasant emotional stimuli upon gastric and intestinal peristalsis. He showed that too much excitement can cause emotional diabetes.

And, gratefully, we accept these demonstrations—of olfactory stimulus-and-response!

In his conclusion, he states his belief that the visceral changes involved in the emotion are determined by the intensity rather than by the quality of the emotion.*

But, as Hunter points out: "Cannon wrongly interprets his facts as antagonistic to the James-Lange theory. On the contrary, they support the theory strongly by indicating a very delicate and widespread bodily disturbance during emotional vagaries." And Perrin comments: "It is certain that Cannon's investigations point to visceral activities as the probable location of emotional response."

As we keep insisting.

---

* A queer distinction—because the intensity depends, largely, upon the quality of the emotion. Fear, for example, has in it an inherent intensity unattainable by the gentler emotions, like benevolence.

Neither Cannon nor Sherrington offers any evidence against smell control. Instead, their findings serve to confirm it. And the logical and introspective criticisms that press so heavily against the old James-Lange theory become supporting buttresses for our sharply specific revision of it.

Cut to a newer pattern, every objection to the theory is answered— but only by using the smell-hypothesis is this result obtained.

Textbooks devote many pages to such matters as: the expressions of the emotions; the origin of emotional expressions; and so on. It would be wandering in foreign fields to discuss them here. The various tests devised, such as the word-association tests, are most of them nonsense. In any case, the sense of smell is never considered, and—as always follows—no conclusive results can be obtained.

I speak of the psychogalvanic reflex merely for the insight it throws into the psychological laboratories which have universally ignored the sense of smell. The galvanometer is an instrument which detects weak electrical currents. Nerve impulses are known to generate electricity. Hence, when the galvanometer is connected by wires with the human body, it presumably responds to neural activity.

I quote: "It has been used successfully, however, only in the case of an emotional response. The resultant is known as the psychogalvanic reflex. The British psychologist Smith compared psychogalvanic measurements with differences in verbal reaction time, and concluded that the galvanometer detects complexes when reaction time fails to do so. . . . On the other hand, the extreme sensitivity of the instrument causes it to respond to many extraneous influences, and therefore interferes with its usefulness in the psychological laboratory."

What other science in the world would discard an instrument because of its sensitivity!

A question often asked me—and one its proponents seem to think unanswerable—is: "Why, if smell is so important, haven't the psychologists found it out, before?"

The answer is: I wonder and wonder, too. But even though dogmatism wears a fine cloak of infallibility, I see no reason—to respect the cloth.

Remember: "The nose is more sensitive than any machine science has devised to test it with." This very sensitivity, it may be, interferes with its usefulness to the research gentry who think they can measure the mind by the rule of thumb.

Here is a law that the impatient science ought some day to adopt: So long as there are outside influences to which even the most delicate instruments will respond, no psychological experiment is of value; and no device which does not respond to such influences may be used in the study of the exquisitely sensitive mechanism of the human mind.

Really, there is hope and promise in that galvanometer!

Behaviorism says that an emotion is an hereditary pattern reaction. To quote Watson: "By pattern reaction, we mean that the separate details of response appear with some constancy, with some regularity, and in approximately the same sequential order each time the exciting stimulus is presented. Stimulus is used broadly to express not only the exciting object but also the surroundings. There is also implied the fact that the general state of the organism must be sensitive (capable of being stimulated to this form of stimulus at the moment)."

Some constancy; *some* regularity; *approximately* the same order; *not only* the exciting object, *but also* the surroundings; *also implied that the organism must be sensitive at the moment!*

We don't hedge that much in this whole book. Watson *has* to hedge! We don't.*

Emotions are smell-responses. These responses vary in the individual with the varying personal odor with which the external odor stimulus must be fused. No part of this requires brainwork. But man has a cerebrum—and uses it—and so there is a further variant: memory.

Here at last the great gray mantle earns our respect. For it never forgets an odor it has been told to remember. Dorsey sees—rather, gets the smell-stimulus of some little girls, and breaks out unconsciously into the nursery rhyme: "Briar, briar, limber lock."

As a child, he is given a raisin containing quinine; eats chocolate and becomes sick. Now, years later, he abhors raisins and chocolate.

I smell musk roses—and I recall an evening ever so long ago. I see moonlight sifting through the vines of a porch, dappling the floor and a girl's white dress with a moving lacework of darkness. I remember how her hair lay in one black ringlet over her shoulder. I hear a train whistle far away; I hear a clock striking somewhere in the house; I hear

---

* The "hereditary" angle wrecks him. Change it to "olfactory"—and the storm-cellar phrases can be left out of the definition.

her voice. And I get a rotten, sick feeling of futility and frustration and incompleteness.

Memory! Yet I never knew, until I smelled a musk-rose, years later, that it was the kind of white flower she wore that night.

Subconscious memory! The great gray storehouse is used by the other senses, sight and hearing. That's quite true.

But always, as Kipling says: "Smells are surer than sights and sounds to make the heartstrings crack."

CHAPTER ELEVEN

CROWD PSYCHOLOGY

"How RELUCTANT," Hudson says, "Nature seems in some cases, to undo her own work! How long she will allow a specialized organ with the correlated instinct, to rest without use yet ready to flash forth on the instant, bright and keen-edged, as in the ancient days of strife, ages past, before peace came to dwell on earth!"

He is referring to the sense of smell. And, in another place, he remarks: "In the Argentine, herds, taken by their odor sense are stricken with panic at times. The whole herd moves at once, avoiding one particular direction. Why? Because the Indians are on the raid—though still, perhaps, thirty miles away—but their odor is perceived by the animals."

Panic! As Kipling says: "Men who trampled each other on the Brooklyn Bridge and knew not why they did it."

An alarm in a Canadian movie-house—and scores of children are trampled to death! Fire in the Iroquois Theatre—and men forget their manhood in mad terror! A sinking ship—and up from the stokehold rush men of the black gang, trampling all in their path, disregarding their officers' pistols, swamping the overcrowded boats!

Five or six years ago, this occurred. A steamer sank in the South Atlantic. The details were hushed up—but not a single man lost his life! Women and children were the ones drowned.

Why do we act so? .... Ancestral memories? .... Fear of death?

How many persons have an actual fear of death? Perhaps you—who read this—have imagination vivid enough, and intelligence clear enough to picture yourself dead clay about to go back to the earth you came from. But—have you ever so imagined yourself? There is an objective envisaging of death. Any potential murderer, any man in anger may picture his enemy dead before him. An avaricious heir may visualize the funeral of his rich kinsman. But, to most persons, death is always something that happens to the other fellow. A devout Christian will face doom with equanimity in the assurance of another life; a true believer may try to die in battle so that Allah will provide him with the musky virgins of Paradise—but at bottom, in neither case, is there a conception of final, personal dissolution—and it is of this that I speak.

Death, as I said, is always something that happens to somebody else. Man has fear of injury, a horror of pain—yes! But that is all.

The fear of death is no explanation of panic.

Ancestral memories? of steamboats? and picture shows? Scarcely. Of the dark? Children, closer to their ancestors, more susceptible to such memories, have if left alone no fear of darkness. The first nine months of the infant's life—before birth—the first few days after— are spent in sightless shadow. Darkness means safety, food, warmth.

What is it, then, that drives men suddenly mad? Fear, of course, but of some dreadful, unknown, *invisible* horror. For fear is bound up with the organism's earliest avoiding movement; it is bound up with chemical irritability; it is bound up in its reception, in its transmission, with the sense of smell.

A herd of deer, unconscious of danger, browses in a mountain meadow upon sweet-scented, flowery grass. With infinite caution, a hunter approaches. One buck lifts his head; gets the first breath of danger—and instantly the whole herd is off in the opposite direction. The other deer haven't scented the hunter; they've simply sensed the fear-odor of the buck.

Emotion means visceral change—and this means: change in body odor. And any strong change in man's individual odor affects not only his sense of smell, but that of every one around him.

I am in the theatre. Stark terror grips me. My terror is a fear signal for those nearby—and a panic is started.

But why was I so terrified? Here is an important point: In every crowd, there are emotionally unstable persons. Perhaps in the hidden past, I have undergone some experience that has left its mark like a deep groove across my mind. Any nerve impulse which strikes the groove follows it. In every other way I am sane—but whenever a message moves along this one groove, I lose sanity. Immediately, I am in the grip of a tremendous fear. . . . and the fear aura that comes from me unconsciously is as unconsciously received by those around.

How can they know about my emotional instability? How can they know that my fear smell is caused by a thought following an old mind groove? The conscious mind may take time to form judgments, but the subconscious mind never does. It *acts*. And panics begin.

At First Manassas, during the Civil War, Union soldiers miles behind the battle-field met frightened men in flight. The fresh soldiers became panic-stricken too; threw away their guns and equipment, and stopped running—some of them—only when they reached Washington city, miles away. History tells of armies—the Saxons with Frederick the Great, for example—twenty-thousand strong, stricken with panic and turning into a disorganized rabble before a single shot was fired.

Why does it take so long to make a trained soldier? Men can learn to march together in two weeks. They can learn to shoot as well as most soldiers do (What were the World War figures? Something like fifty thousand bullets fired for every man killed,) in less time than that—but America let its troops enter the trenches only after a year's training . . . Why? Because a tremendous amount of field-work is required in learning to resist the fear-contagion. Ninety percent of a soldier's training is the formation of a conditioned reflex against subconscious smell.

And fear is not the only crowd-affecting odor. Actors will tell you that emotions move in waves through a theatre. A Fourth of July speaker knows how reactions travel in widening circles through a crowd. Always there are centers—persons who feel the emotions first—and from these centers the contagion spreads. An evangelist playing on and for emotional response, like Billy Sunday or Aimee the well-beloved, always sees to it that there are good, dependable, emotionally unstable sisters and brothers scattered throughout the temple. And these act as leaveners for the mass.

Lafcadio Hearn was interested in crowd movement—in the way pedestrians intuitively avoid each other. He wrote: "My knowledge is certainly intuitive—a mere matter of feeling—and I know not with what to compare it—except that blind faculty by which, in absolute darkness, one becomes aware of the proximity of bulky articles without touching them. . . . Furthermore, I find that whenever automatic—or at least semi-conscious action—is replaced by reasoned action, I always blunder."

"Indeed my personal experience has convinced me that what pilots me quickly and safely through a thick mass is not conscious observation at all, but unreasoning, intuitive perception."

And, he concludes: "This quality of intuitive self-direction in a crowd is shared by men with very inferior forms of animal being."

As smell is!

CHAPTER TWELVE

FRIENDSHIPS AND ANTIPATHIES

INSTINCTIVELY you like a man, or hate him. He may be the greatest scoundrel unhanged. You can't help liking him. He may be a good man, an honorable man, a decent man. You avoid him as if he were a snake. Why?

The question remains one of the deeper pitfalls to psychology. The Freudians blame it on sex—but the Freudians find sex at the bottom of everything. The behaviorists say it is the result of adolescent reaction to authority, *and* sexual stimuli bringing out emotional reactions. Watson puts it on a "faute de mieux" basis, and suggests that if you don't like his reasoning, try and do better! He's somewhat aggrieved about it. You get the impression that he isn't satisfied with his own explanation—and would like to ignore the whole matter.

He can't, of course. The impression you make on other persons is so very important. Your true character, your real personality matter only to yourself.

You meet a Lightnin' Bill Jones—and you find yourself liking the old no-'count idler. You know he's a cumberer of the earth; you know he's not worth the powder it would take to blow him to where he's inevitably going; he's not handsome; he has no social graces; his voice isn't any lyric tenor. Why do you prefer him to a better man?

The books list a row of reasons as long as your arm. Check them off one by one—and you'll discover that old Lightnin' hasn't a single favorable quality. He's all liabilities and no assets.

He doesn't remind you in the least of your father or your teacher or any man you knew in your youth. He doesn't look like your mother. And you—a fairly normal individual who's never been "psyched" up to now—doubt whether you ever owned an Oedipus complex, anyway. You're reasonably sure of your unmixed masculinity—and you reject the suggestion of a sex-motif, analyzed or unanalyzed.

All you're sure of is that you like Lightnin' Bill Jones. It isn't a matter of judgment. Good sense tells you you ought not to like him. Your eyes tell you that he has the appearance—and lack of movement—of a disreputable old vagabond. Your ears tell you that he mistreats the English language in a way that's not even funny. But—you know he's got an attractive personality.

How do you know? By sight? What's attractive about his appearance! By sound? You've just finished thinking that you'd like him even better if he'd keep his mouth shut.

"All our knowledge," Gassendi told Descartes, "comes to us through our senses." Apply that law here! You've three distance receptors. Two of them, eye and ear, don't care for old Lightnin'. The third—*must!*

Your nose approves of Lightnin'. Your sense of smell is subconsciously at work once more.

And that good pious brother who's just walked away—and by doing so took away that vague feeling of discomfort you were experiencing—your nose didn't care for him. He could endow nineteen more churches —and still your nose wouldn't like him. Nor would the rest of you, either.

Eyesight helps; hearing helps—and there's always the skin you love to touch. But when all's said and done, when the nose breathes rapturously: "Lord, what a nice, sweet smell!" the other senses haven't a chance to offer their opinions. And if the nose says: "Take it away!" it's a case of

> "I do not like thee, Doctor Fell,
> Why it is, I cannot tell."

You had a nose long before you had an acquisitive instinct, or a respect for property, or a slavish adoration of the Social Register. You had a nose to tell you of strangeness or familiarity; of danger or safety; of antipathy or liking. And the nose hasn't changed in the least. It's a democratic sense, unaffected by glitter, scornful of Paris gowns or shimmering jewels or the rustle of silk. To an extent, perfumes can fool it, perhaps, but nothing else can. And—it speaks first—and loudest!

First impressions! They're olfactory ones, at bottom. Your nose tests each newcomer, and gives you its opinion. And your nose is always right.

First impressions are frequently wrong, you say? They're never wrong. It's the conclusions drawn from them that are wrong. You meet a man. You like him. Perhaps you put trust in him—and he robs you of your pocketbook, or forges your name to a check.

"Such an attractive scoundrel!" you mourn. "It just goes to show— never trust first impressions!"

Your nose does not know you have a pocketbook. It has never heard of a check—and it's interested in neither. Your nose is concerned with your body inside your skin—not with your personal property outside it. Your nose passes automatic judgment on every man you meet— and the instinctive feeling of friendship or dislike is the result.

If you are a man, and you happen to meet a woman, you are predisposed to like her. She has, always, the feminine odor which appeals to you, and so long as she has not doused herself with certain sharply masculine sex-smells disguised as perfumes, she keeps this feminine appeal.

If you are a young man—unmarried, or unhappily married—you experience, at each first meeting with a woman, a sharp sensation of curiosity, of inquiry, as your nose seeks to determine whether this is the right woman or not. I remember once, years ago, I sat in an arbor, reading. Something made me look up, suddenly . . . . and I saw a girl, and immediately felt a mental thud! of disappointment. Whenever I met the girl, afterwards, I felt the same strange fall-in-spirits. The thing was ridiculous! It puzzled me every time I thought about it.

Now I know the explanation. It was simply that she was almost—but not quite . . . Her chemistry was so close, so very close—but my nose detected some small strangeness and told me so.

We must not forget that the sense of smell, more than any other sense, *remembers*. And it is by this memory, I think, that we can grant most of what the Behaviorists and Freudians have asserted.

A child remembers his father, pleasantly. The child, grown up, will have a friendly attitude toward any man who reminds him of the father. A boy loves his mother; may in later years marry a girl who reminds him of this mother. These statements are unquestionably true. But—the reminiscences are preponderantly olfactory, and the likenesses preponderantly chemical.

A boy doesn't marry a girl who *looks* like his mother. He doesn't make friends with men who *look* like his father. He depends on his nose to detect the likenesses.

With that understood, how the mystery resolves itself! No need now to make every man more or less homosexual; no need to change manfriendship into unanalyzed sexual love; no need of Libidines and Oedipus-complexes; no need to aver that grown men are still afraid of the sound of their teacher's voice!

Two good men meet a third—and one feels an instant liking for, one feels an instant antipathy toward the stranger. The explanation seems simple enough: There are three components in the total smellmessages the men receive, and of these only one, the objective stimulus, is the same. But certainly, the individual, personal-odor components are different—and the associated memories each man gets from that particular objective odor—or odors similar to it—are different, too. Why shouldn't there be different reactions!

Zwaardemaker had our answer in his grasp, though he didn't know it. "Olfactory sensations," he said, "awake vague and half-understood perceptions which are accompanied by strong emotions. The emotion dominates us—but the sensation which was the cause of it, remains unperceived."

Vague and half-understood perceptions! "Something told me." . . . "Somehow I knew." . . . "By instinct." . . . "Instinctively." . . . "Intuitively." . . . "Intuition told me." . . . "It struck me." . . . "All of a sudden, I knew." . . . "I sensed." . . . "I felt his personality." . . . "I was aware." . . . "It seemed to me." . . . "I felt." . . .

Notice the words you use. You *know* it's not sight or sound. You *know* there's something else. . . . And I think now you know what that something else is.

When you speak of the personality of an individual, what adjectives do you use? "Strong," "compelling," "attractive," "pleasing," "forceful"—or "weak," or "lacking," or "vague."

Unconsciously, intuitively, you use words that describe—a smell!

This is peculiar and significant. Instinctively we differentiate almost always between undoubted visual impressions and shadowy olfactory ones. When we say: "I saw," and tell of something definitely seen, we speak of a visual image. But when we say: "It struck me," "It seemed to me," we actually mean that the nose has received a message.

At such times, we're never able to give a logical explanation of our suddenly formed opinion. We can only achieve a faltering: "I thought . . . Why? . . . I don't know. . . . It just seemed . . . ." And so on. We *know* a sensation came to us—yet none of the accredited senses seems to be responsible.

And—nine times out of ten, when we say: "I felt," we don't mean that we've used the sense of touch!

One girl's the belle of the ball; another's a wall-flower. One has sex-appeal; another hasn't. One man's popular. Another knows the five-foot book shelf by heart and still nobody likes him. One man's a lady's darling and an interior decorator. Another's a prince in a locker room and a plague in a lady's boudoir. The town beauty falls in love with the town beast. Cophetua chooses the beggar maid.

It's *Personality!*

Lightnin' Bill Jones has it. So has everyone else, whether it be attractive, repellent, or in between. And each man's is different; and each
woman's.

As different as the individual odor that personality actually is!

There's a Tartar in Tolstoi's "Prisoner of the Caucasus," who sees
around the head of every innocent Russian the malignant faces of the
Russian soldiers who murdered six of his sons and forced him to kill
the seventh with his own hands. The old fellow bristles and snarls in
fury at sight of every Russian. It's hate at first sight, always. . . .

. . . . Only, it isn't the *sight* of them that maddens him.

Race odor has caused more wars than religion. The French personality—the German personality—what are they to each other but odors
of strangeness!

Some day, when humanity realizes the absurdity of fighting over a
smell—wars will cease.

I remember, thirty years ago, I met some new neighbors for the first
time. I met them—and commented on them at home.

"Mr. and Mrs. Tompkins," I told my family, "certainly look alike.
You can tell they're kin to each other."

My family laughed.

I didn't understand. It seemed to me—seven years old— that persons living in the same house were always blood-kin. I had a vague belief that my mother and father were more closely akin than my sister
and I were. And—I was very sure that Mr. and Mrs. Tompkins were
alike.

My family explained. I went down the street and stared at the new
neighbors again. And, to my horror, they did not look alike at all. Mr.
Tompkins was fat and President-Hoover-ish; Mrs. Tompkins was thin
and acidulous. The man had a face like a flattened full moon from
which two piglike blue eyes peeped; the lady's face was built around
a tremendous beak, and her eyes were fierce as a hawk's.

The new neighbors did not look alike in the least—but still, they
were alike. I couldn't tell how, but I was even more confirmed in my
belief than before. Their alikeness remained one of the enigmas of life.
Every few years I'd take it out, and turn it around—and try to puzzle
out the answer.

Now I know. Their personalities were alike—and my nose caught
the resemblance. Perhaps living together, they gave off a common nest-

odor, as ants do; or eating the same food may have had some effect . . . for all I can remember of the Tompkins now are how they looked and the fact that the lady made home-made bread every day, and her house was redolent with the baking.

I seem to have neglected the matter of food smells anyway in this book. It may be: race and national odor are caused by race and national food, and the German's sauerkraut, the Frenchman's bouillabaisse have something to do with their racial antipathy. If so, a proper way to end war would be to exchange cooks—and cook books. Perhaps drinks are of importance, too—and if that is so, no wonder Prohibition-America, during the 'twenties, didn't have a friend in the world!

CHAPTER THIRTEEN

LOVE

I HAVE BEEN somewhat hysterically urged by almost everybody to omit this chapter. I am told—the taboo is working overtime, of course—that there is something revolting about our idea—as an explanation of love, at least.

And so I show that there *is* something sacred—still—and save my notes for the next chapter—which treats of Sex.

Of love, I say nothing.*

---

* Maybe I shouldn't quote, either, but—

"Henry IV's passion for the beautiful Gabrielle originated at the instant when, at a ball, he wiped his brow with her handkerchief."

"Henry III, at the betrothal feast of the King of Navarre and Margaret of Valois, accidentally dried his face with a garment of Margaret of Cleves, moist with her perspiration. Although she was the bride of the Prince of Conde, he immediately conceived so violent a passion for her that he afterwards, as history shows, made her life very unhappy."

"In Tolstoi's *War and Peace,* Count Peter suddenly decides to marry Princess Helene —after inhaling her odor at a ball."

And—

"George Meredith's sentimental English youth fell out of love with Vittoria, the heroic Italian singer, because he detected a whiff of tobacco smoke in her hair."

CHAPTER FOURTEEN

SEX

FOOD OR POISON—shall I eat it or leave it alone? Friend or foe—shall I stand still or run?

In a unisexual world, these are the two great problems of life. But when Sex appears, another question presents itself to become, at certain periods, the most important query of all.

Man sometimes tries to answer it by plucking petals from a flower: "She loves me . . . she loves me not . . . she loves me." Lower animals have simpler methods.

Love to them is a forthright question of sex attraction—as it was to man until seven hundred years ago. And as it is now—straitly, to most men. The courts of love, the chivalrous notions of love, are fine catchwords to catch readers for romantic novels—but if you put one boy and one girl together in a dark place and stir, you'll get an elemental mixture.

In this question of Sex, I find I have far too many notes. Every twist and perversion has been card-indexed and catalogued. Every possible cause, including the sense of smell, has been named. Every neurosis has been attributed to it; all the Unconscious, Freud says, has come from it. And it *is* important. There are times when it is the only drive —times when food and fear are forgotten. It is made attractive because it is so necessary to life; it has the most delicate balance of all the bodily functions; it is the most easily unbalanced; and when so, can have the most devastating results. We are told that the sublimation of it causes creative art and religion. We are told that the degradation of it makes us worse than beasts. We are told . . . the X marks nine thousand different things I omit.

Man, in his sex life, differs markedly from most other animals, because man is *always* interested in sex. This is unnatural. And because the twist has rooted itself in man's very being, we note its effect in every manifestation of human temperament and human behavior. Among adults, it is difficult to conceive of a situation in which the sex angle does not enter.

But after all is said and done, Sex isn't the whole works. There are other things equally important in their way, equally as inextricably a part of the individual's personality. A psychoanalysis that does not consider these other things is trying to explain a part as the whole.

Moss, in his study of animal drives, found that both food- and sex-hunger were stronger than punishment. But—he found sex was inferior to hunger in driving force, while maternal love was weaker than either.

Watson says: "A young man may be extremely sensitive to the blandishments of every female he meets, while in the unmarried state, and may show considerable excitement and over-reaction on such occasions. Usually he becomes considerably less sensitive after being happily married."

Professor Watson is some several years removed from young manhood. Otherwise, he wouldn't qualify his statement so. A young man *is* (and no "may be" about it!) extremely sensitive to every young female he meets—whether she uses blandishments or not. And if he is even fairly happily married, he becomes—so far as a member of the sex can —completely insensitive. . . . Otherwise, so much dictation couldn't be accomplished in offices.

I think we can consider Sex here—so far as it concerns our theory—without diving into sewage, and bringing up particularly nasty bits triumphantly, as the Freudians do.

First of all: we have no need for any sex life in babies; no belief in unanalyzed sexual instincts in children under five. The baby regards its mother as food and protection only. The mother-odor means these things to it and no more. The father odor means protection always. The Unconscious is no repository of smutty stories—and there is no sexual Libido.

We must remember that after all, the child has a very weak sense of smell. Hudson misinterpreted the facts. The child receives more *conscious* odor-messages, has more *conscious* sensations—because the taboo is not so strongly re-established. But no child has the potential olfactory acuity of the adult.

For here is a very curious thing: the only sense which changes at puberty is: smell. The supreme upheaval that occurs when girl-child becomes woman; when boy-child becomes man—is disregarded completely by the senses of sight and hearing. The child sees, hears, as well as the adult.

But at puberty, along with the development of secondary sex characteristics, there is a tremendous increase in the size of the nasal septum. There is a marked increase in the olfactory area and in olfactory acuity. And, for the first time, there is a sex-difference in the attraction of various odors. Body-odor changes too, for certain sex-odors become components of it for the first time. Sensitiveness to body odors becomes more marked—for body odors, like perfumes, are nervous stimulants.

Gustav Jager regards the sexual instinct as mainly or altogether olfactory. Personal odors, Havelock Ellis says, cannot be a matter of in-
difference in the most intimate of all relationships. He records the fact
that many neurasthenics—particularly those sexually so—are peculiarly
susceptible to olfactory influences. He speaks of the association of
smell and sexuality which is observable throughout the whole animal
world. A sympathy, he says, certainly exists between the olfactory and
sexual regions. There is a very intimate relation, both in men and women, between the olfactory mucous membrane of the nose and the
whole genital apparatus. They frequently show a sympathetic action. Influences acting on the genital sphere will affect the nose, and occasion-
ally, it is probable, influences acting on the nose reflexly influence the
genital sphere. . . . And he speaks of the popular expression: "Bride's
cold," as being often true.

Further evidence of the tie-in is that to the commoner forms of per-
verts—Lesbians and male homosexuals—the olfactory area of the nose
becomes the realm of sexual satisfaction.

I have been more-or-less fortifying myself in the last few pages for
this:

You fall in love. If you are asked Why? you answer with a catalogue
of the adored one's good points. Isn't it true that you discover these—
afterwards? To "make sense" out of the sensory thunderbolt that struck
you?

Perrin says that the lover, dominated by his emotion, cannot associate
unpleasant or unlovely things with the object of his emotion.

This is absurd. The lover, when *with* the girl cannot associate such
thoughts with her, because he is dominated by her—sensuously. But—
away from her, he certainly can. He is troubled by doubts, jealousies,
suspicions—all unpleasant, unlovely things.

And the reason is: away from her, other odors come to him. He is
happy only while with her, because it is solely then that he feels in an
olfactory heaven. His subconscious mind demands more and more of
this soul-stirring happiness . . . and he proposes to the girl. If *her* nose
says: "Yes,"—and no civilized factors intervene—they are married.

And here, I think, is the real explanation of unhappy wedlock and
divorce. It is the entry—disturbing to Nature—of unnatural factors.
And, like Adam in the matter of an apple—I blame the woman, most.
Marriage, to most men at least, is a question of the girl; marriage to
most women is unduly concerned with property values and prospects.

Take any two women discussing the marriage of a third—and the first question asked and answered is: "Did she make a good marriage?" Shylock would be interested in the goodness meant.

Most men select their wives instinctively—in other words, by the natural aid of the sense of smell. Most women prefer a more business-like method—with free use of Dunn and Bradstreet.

Man is simple in his seeking. He finds the answer to his mate-hunger —and proposes. Woman tries to solve all her life-problems at once: food and fear and sex. Oftentimes, the man who can assure her against famine and worldly difficulties is barely tolerable sexually.

Her proper mate is a boy of her own age, not yet able to support himself. But she dismisses him, and marries a good provider, who can provide for her—by court order—even after he becomes sexually abhorrent to her. The childhood sweetheart, in turn, wastes himself and his health and his youth in devious civilized ways, until he is old enough, and wealthy enough, to be considered a good provider by a girl of the next half-generation.

There's this side to it: If civilization, and morals, and family life, and laws are man-made institutions—and man, somewhat wonderingly, takes credit for them—then the fault is, at bottom, his. For these things have been as prisoning walls to woman; and so it's little wonder that our grandmother usually picked the jailor-husband who'd give her the finest, most comfortable cell. After all, in those days a girl's scope was so limited; her field for husband-hunting so small—just the family parlor, and what few candidates happened in. And each one might be the last—and there was that spectre of old-maidhood to frighten her on. . . . Maybe Grandma wasn't so blameworthy, at that. She didn't have free choice—or a fair chance.

Today, though, the situation is different. Woman has gone into business, into sports, into the world. Where yesterday she knew, and could choose from, half a dozen men, perhaps; today her acquaintance, college, social, business, may run into the hundreds. Out of so many, she's almost sure to meet the right man. And if, because of some old-fashioned, old-woman's advice, she does not accept him, it is her own fault. But because there is a thrill in the right man—like the thrill of the one woman—stronger than any other sensation a human experiences throughout life, I doubt whether old-fashioned requirements for a husband will seem so important in the future. I think that even in marriage —the mating with which olfaction is primarily concerned; in which, among women, it has heretofore received so little consideration—the sense of smell will come into its own.

I am far from predicting that divorces will end. The right man—the right woman—may not be that way, always. Personal odor changes, and personal attraction follows. But—divorce court judges will be graduate doctors—not lawyers.

Civilization is not man's proper biologic state. The nose selects instinctively, with little regard for the conditioned reflexes—the habits—civilization has taught us. The nose says: "Here, physiologically, is your mate." On a desert island, in a state of barbarism, the marriage would be one of those fabulous ones, made in heaven. On earth, sometimes, it seems to have had another country of origin.

Even so, conditions may be improved tremendously, simply by a recognition of the real nature of marital differences.

"He's not the man I married!" It's quite possible. Some change in his bodily chemistry may have altered the message your nose receives. But on the other hand, he may be! The change may be in yourself—and it is this change that makes the resultant smell message sent to your brain, repellent.

"I try every way I can to please him, but he acts as if he can't stand me!" He can't. It's not your fault, or his. It's a straight-out matter of chemistry.

"He met that woman—and she's all he thinks about." He's not to blame. Probably he tries, with all his conscious mind, to stay away from her. But he's like a led dog. His subconscious mind controls him—through his subconscious, chemical sense.

Once this is recognized, we'll quit blaming men and women for things they can not help. We'll realize that the holier-than-thou attitude means that its owner has been lucky so far, and nothing else.

I think we'll quit censuring entirely. We'll cure.

There's a way.

CHAPTER FIFTEEN

MOODS

THE SAVAGES burn down the village when anyone dies. It's too bad we can't do the same.

Death must have such a pervasive, powerful odor. I don't like cats. I've seen them come in packs whenever a man died. I've watched them at windows, at doors, trying to get to the corpse. And once I saw a man, fond of the beasts, who had died alone in a house with his pets. And since then, I've been willing to throw twentieth century knowledge aside and agree with the superstitious ones who say cats are witches in disguise. For if ever I saw mad, supernatural, horrible creatures, I saw them then.

The crows in Cornwall, we learn from NATURE, gather on the roof of a house where a man is dying. Pliny says such birds smell the odor of approaching dissolution. Medical books speak of the smell of death. Vultures scent a body miles and miles away. I remember watching French soldiers bury their dead—scores of them in a long trench. . . . And I remember how the memory lasted.

Cats tell us most—and what a fearful thing the smell of death must be! Unconsciously we smell it; unconsciously we are affected by it. It lingers—and its effect lingers, too.

The savage burns his village and gets rid of it. We live in stone houses—and the memory lingers on.

With other odors! I open an old jewel box belonging to my sainted grandmother. Immediately the old lady stands before me. The associative power of odors! The forgotten memories they evoke! I feel a mood of sadness because she is no longer alive.

My mother dies. I live in the house she died in. I visit her room. I treasure all the souvenirs of her I can find. I am inconsolable. The bright world is grown gray to me. I care for nothing; I want nothing; I live in a sad house of memory.

Why? Nothing I can do can help her. Nothing will bring her back. Why do I grieve on—and on—and on?

I am a victim of melancholia, my friends say.

Melancholia?

All around me are the faint odors I associate with my loss. Unconsciously, continuously, they affect me; make me lose health and sanity.

The savages knew better. They burnt their villages—and got rid of odor-stimuli.

Till most recently, it was quite the proper thing for a woman to go to a hospital to give birth to her child. Then some doctors noticed that women, at home, under guarded, aseptic conditions, seemed to do better. There was something disturbing about hospital air.

Of course there was! The odor of sickness, of suffering humanity.

Hospitals are certainly wrong, in principle. A sick man gets sicker the moment he enters, or is carried through the doors. The use of disinfectants and deodorants (affecting the trigeminal nerve) can have no effect whatever on the stimuli unconsciously received by the olfactory nerves, and these sickroom stimuli are bound to be depressing ones. As such, they militate against treatment, delay recovery—for every patient, so long as he is in the place, feels low-spirited, depressed, under the sway of the dreary hospital mood.

A doctor prescribes an ocean voyage, a change of scene for a client who is listlessly convalescent. The change of scene works wonders. Why shouldn't it! It means a change of environmental odors, too—and the old ones were the direct cause of the sick man's listless mood.

For any mood caused by objective stimuli can be changed by a change in these stimuli. Even when it is a result of visceral odors, inwardly received, it may be lightened by external change. For there seems to be a sort of unconscious substitution.

Visceral upflarings (such as those due to rage) subside to normal very slowly—and meantime we may have come to associate the (actually internal) anger scent with some specific odor reaching us from our environment. After all, the composite odor is the stimulus the sense of smell reacts to; and our consciousness is always of external objects, and our ears and eyes are always at work suggesting possible outside sources of the continued emotion.

My banker refuses me a loan. I go home and beat my wife—who spanks the baby.

It's all reasonable enough. It's just too bad the baby can't go out and bite the banker—and complete the circle.

Certain odors may have an associative poignancy, may have grooved their emotional channels across the brain. In this case, whenever the odor is received, the mood is regained.

As examples of this: we read of a man who was always insanely ill-tempered in the dining-room—because he had an unconscious antipathy to the scent of horse-radish. It was always on the table; it always affected him—but he always found fault with his family. We are told of a man who classified all unpleasant odors as the smell of honey. An investigator found that the man had been made violently sick by eating honey, some weeks before.

Pleasant moods run the same course as unpleasant ones. The girl says "Yes," and I float about my business on a magic carpet of delight. Word comes to me that my partner has absconded with the partnership's funds. I'm not interested. I'm told that my house has burned down, just after the insurance lapsed—and I say that I never liked the old place, anyway. I am too happy to be bothered about anything. Why? . . . The girl said "Yes."

And probably, in addition, some of that divine emanation that makes me love her still clings to my coat and sleeves and moustache—mementoes of what happened after she said the wondrous word. *And it is this that causes my happiness to continue!*

"According to one view, an individual's temperament is determined by his most characteristic mood or moods." (Perrin). Let's make this more definite:

The most characteristic moods are caused (1) by specific stimuli from the environment in which the individual passes most of his time, and (2), by the sensitiveness of the individual to these stimuli. This sensitiveness is determined by and varies with the urgency and importance of interoceptive odor-messages.

For here is a point that needs stressing: in normal, untroubled times, personal odor is translated into a sense of vague well-being. In this state, the individual pays little attention to himself, and much to the world about him.

Inherently, though, man is more interested in himself than in the outward world. So long as things are working smoothly within him, he responds to his environment. But when things go wrong inside, outside things matter little. A very sick man doesn't care *who* is elected President. The only external stimuli to which he responds are those which confirm or strengthen the mood he is already in. Even slight noises irritate him tremendously; lights are too bright; the mattress is uncomfortable. Bad news affords him gloomy satisfaction; but joyful news is received joylessly.

Temperament is a matter of body condition first and most importantly. Sometime—a long time—later, it is affected by environment.

And so we go back to the Greeks—and find that Galen, who spoke of the humors causing temperament, was—as far as his knowledge went—absolutely right.

"The world of our environment keeps beating in upon us as stimuli —pressures, chemical substances, sound waves, light waves."

And again: "We are earthbound within the limits set by our physical sensory equipment; nor can our thinking transcend the realm of sense experience."

Dorsey says one; quotes the other. *Sensations are from the senses.* That's dogma—and more than dogma. It's natural law. And it means: that in the exercise of every human faculty, some sense (or senses) has given the Go-signal; that if one sense isn't responsible, another (or others) must be. It means: that if eye- and ear-reactions are inadequate to explain observed phenomena, and smell—if functioning—will explain them—*then there must be such a functioning smell-sense!*

So—we go on to a faculty we can exercise in absolute darkness—in a sound-proofed room.

"A sense arises," says Havelock Ellis, "vague, but usually recognizable, of something seen casually long years ago, and possibly never thought of since, and possessing no kind of known association either with the matter in hand or with my personal life generally. It comes to the surface as softly, as unexpectedly, as disconnectedly, as a minute bubble might arise and break on the surface of an actual stream from ancient organic material silently disintegrating in the depths below. . . . Every one who has travelled much cannot fail to possess, hidden in his psychic depths—a practically infinite number of such forgotten pictures, devoid of all personal emotion. It is possible to maintain, as a matter of theory, that when they come up to consciousness, they are evoked by some real, though untraceable resemblance to the psychic or physical state existing when they reappear. But that theory cannot be demonstrated."

Such pictures as the musk rose brings to me—without the bitterness; or my grandmother's jewel-box—without the momentary sadness.

But Havelock Ellis is speaking of day dreams—the pictures that float disconnectedly, one by one, out of the Subconscious into consciousness. It is significant that he uses the simile of breaking gas-bubbles—for these things *are* odor-stimuli, bringing up associated memories from the past.

Disconnectedly? Of course! A passing truck carries apples from the Albermarle. Faint odor is wafted through the window—and I see the Blue Ridge country in late summer, with a smoky haze upon the higher hills. A peddler trundles his pushcart of bananas—and I see lush tropic shores, with a blue sea beating—beating. An Englishman passes, baggy in his Harris tweeds—and the peaty odor brings its picture of thatched cabin and open fireplace.

Softly, unexpectedly—they *must* come so, as the outer world sends its varied messages

Ellis says it is possible to maintain—as a matter of theory—that they are evoked by some real though untraceable resemblance to the psychic or physical state existing when they reappear.

What sense *can* bring real but untraceable messages? Any resemblance of form or color would be instantly apparent. Any familiar sound would be recognized. Faint odors, though, subconsciously evoking memories of similar odors, subconsciously received in the past—are equally real and equally (tabooed as we are) untraceable.

Gissing, the novelist, in *The Private Papers of Henry Ryecroft,* says of day dreams: "A thought, a phrase, an odor, a touch, a posture of the body may possibly have furnished the link of association."

Consider: Thought, phrase, touch, posture—these would be identifiable. A subconscious odor—would not!

If we are earthbound within the limits of our physical sensory equipment—and we are!—we must explain day dreams by our physical senses —and the only one which will explain them is: the sense of smell.

Havelock Ellis speaks of hypnagogic paramnesia—a terrifying phrase for the half-awake form of what is sometimes termed pseudo-reminiscence. Dickens experienced it—and described it in *David Copperfield*: "A feeling of what we are saying or doing having been said or done before—in a remote time; of having been surrounded, dim ages past, by the same faces, objects and circumstances, of our knowing perfectly what will be said next, as if we suddenly remembered it."

"Probably," Ellis comments, "the idea of the pre-existence and transmigration of souls comes from this. In any case, the subject is liable to an emotion of distress which would scarcely be caused by the coincidence of resemblance with a real previous experience. . . . Laland found this sort of paramnesia in thirty out of one hundred people. Heymans found it in a considerable proportion of students of both sexes whom he tested. The experience seems to affect educated people and notably people of more than average intellect—including many artists and emotional workers. Shelley had it.

"Many psychologists maintain that these pseudo-reminiscences indicate a real but confused memory of past events in our present life—dim recollections which the subject is totally unable to locate."

It is noteworthy that artificial anesthesia by drugs which produce an abnormal sleep—also produces paramnesia.

Grasset considers that the phenomenon is due to a subconscious impression previously received, but only reaching consciousness under the influence of the new, similar impression.

> "Like something I remember
> A great while since—a long, long time ago."

All this fits in with our smell-hypothesis. However (and I bear up under your laughter), because of certain strictly personal occurrences, I exercise an idiosyncracy—and in this one case, doubt the connection.

CHAPTER SIXTEEN

DREAMS

SOMEWHERE in our hypothesis, it seems to me, there should lie an adequate explanation of sleep—but I have slept on the problem some several nights, and I'm still groping.

It may be because sleep is an attribute of the highest centers of consciousness—those of learned control and association of motor and speech mechanisms, which Dorsey says are the most recent acquisitions to the nervous system, the least organized at birth, the most modifiable from birth. Certainly with these olfaction has little concern. But I can't get it out of my mind that I am overlooking something.

Anyway, there is nothing in the matter to do great damage to our hypothesis.

For the sense of smell never sleeps.

Dreams, at least, we can explain—and not in the Freudian way of symbolic fulfillments of repressed sexual motives. There is no relationship whatever between dreams and any sort of motives.

Freud says that orthodox psychologists have never been able to explain the things. "Dreams, according to them, were nothing other than the result of physiological sensation, originating from unequal soundness of sleep in different parts of the brain."

And, he comments: "Any psychology unable to explain dreams cannot expect to be recognized as a science."

He explains them thus: "Dreams are residues of waking life, but the one which stands out—the isolated thought—is found to be an impulse in the form of a wish, often of a repellent kind, foreign to the waking life of a patient and disavowed by him with indignation.

"Some of the Ego remains in the form of a censorship of dreams, and forbids the unconscious impulse to express itself in the form which it would properly assume. In consequence of this censorship the latent dream thoughts are obliged to submit to being altered unrecognizably.

"A dream is the (disguised) fulfillment of a (repressed) wish.

"If the meaning of the dream becomes too plain, the sleeper awakes in terror. (anxiety dreams).

"A similar failure in the function of dreaming occurs if outside stimuli are too strong to be warded off. (awakening dreams).

"Hunger, thirst, or the need to excrete can cause dreams of satisfaction."

If Freud could prove that humans are kin to angels—of whom we know nothing—instead of animals, of which we know much—he would have a much better case. For, so far as the students of their behavior can determine, animals dream.

My dog seems to be dreaming now. And—I wonder if her dream is an impulse in the form of a wish, often of a repellent kind, foreign to her waking life, and disavowed by her with indignation?

Over and over again, Lloyd Morgan's Canon of animal behavior interferes with the psychologists' use of big words.

Let's wander through Havelock Ellis' *World of Dreams*. He speaks of the two forms of attention: voluntary, and spontaneous.

"Voluntary attention is the product of education and training. It is directed by extrinsic force, is the result of deliberation, and is accompanied by some feeling of effort. It always acts in the muscles and by the muscles. . . . Without muscular tension there can be no voluntary attention.

"Spontaneous attention is that more fundamental kind of attention which exists anteriorly to any education or training, and is the only kind of attention which animals and young children are capable of. It may be weak or strong, but always and everywhere it is based on the emotional state. These two kinds of attention are at the opposite poles from each other, and are incompatible with each other. There can be no doubt that it is voluntary attention that is defective (though it may not be entirely absent) in dreams. But all the characters of spontaneous attention are present. The attention we exercise in dreams is mainly of this fundamental, automatic, involuntary character, conditioned by the emotions we experience and for the most part escaping all the efforts of our voluntary attention."

—And what does "spontaneous attention of a fundamental, automatic, involuntary character, conditioned by the emotions" mean to us now!

Dreams, according to Ellis, are:

A: REPRESENTATIVE—connected through the fact of association with the waking life of the past.

    1. Recent: Belonging to the previous day.

    2. Old: From experience of many years past, frequently from early youth.

B: PRESENTATIVE—connected through some excitation with the immediate present.

    1. According as they refer to external stimuli present to the senses.

    2. According as they refer to internal disturbances within the organism.

Any or all of these four sub-groups, he adds, may become woven together in the same dream . . . and into these divisions the whole of our dream life may be analyzed.

He continues: "Whether the influences which stimulate our dreams arise from without or from within our organism, they are always filtered and diffused through the obscured channels of perception. They reach the brain at least in a vague and massive shape that may or may not betray to waking analysis the source from which they arise, but will certainly have become so changed in these organic channels that their affective tone will be predominant. They are, that is to say, largely transformed into *emotion*. And when so transformed they become the origin of what we regard as the imaginative element in dreams. . . .

". . . While the excessiveness of dream imagery is mainly due to the conditions of the nervous sensory and motor channels, there is also a heightened affectability of the cerebral centers themselves perhaps due to their state of dissociation or absence of apperception, which leads us in our dreams to react extravagantly to the stimuli that reach the brain. . . .

". . . All these various considerations lead us to a central fact in the psychology of dreaming, the controlling power of emotion on dream ideas. From our present point of view we are now able to say that the chief function of dreams is to supply adequate theories to account for the magnified emotional impulses which are borne in a sleeping consciousness . . . Sleeping consciousness is assailed by waves of emotion from various parts of the organism, but is entirely unable to detect their origin, and therefore invents an explanation of them. . . . The fundamental source of our dream life may thus be said to be emotion.

"This has been recognized by Herrick, McDougall (who considers dreaming a succession of intense states of feeling supported by a minimum of ideational content; the feeling is primary); Grace Andrews, who found that dream emotions are often stronger and more vivid than those of waking life; P. Meunier: 'The substratum of a dream consists of an emotional state.' "

"Our dreams," Ellis goes on, "bear witness to the fact that our intelligence is often but a tool in the hands of our emotions. In this we see the basis of the symbolism which plays so real a part in dreams. It rests on the fact that we associate two things—even if the one happens to be physical and the other mental—which both happen to imply a similar state of feeling. Symbolism of this kind is indeed characteristic of the human mind at all times—in all stages of its development."

"The dream state" (he says it so much better than I can that I continue quoting), "represents the normal psychic state of childhood; sleeping consciousness the embryonic psychic state.... Sleeping consciousness is the primitive embryonic consciousness. This is indicated, it seems to me, by the fact that in many animals the embryonic position is the position of rest and sleep. Ducklings and chicks in the shell have their heads beneath their wings. The dog lies with his feet together, head flexed, hind quarters drawn up. Man, alike in the womb and asleep, tends to be curled up, with the flexors predominating over the extensors."

Here's Sully's opinion: "In sleep we have a reversion to a more primitive type of consciousness."

"Dreaming," Jastrow contributes, "may be viewed as a reversion to a more primitive type of thought."

Finally, Bergson: "The dream state is the substratum of our normal life. Nothing is added in waking life; on the contrary, waking life is obtained by the limitation, concentration, and tension of that diffused psychological life which is the life of dreaming. To be awake is to eliminate, to choose, to concentrate the totality of the diffused life of dreaming to a point, to a practical problem. To be awake is to will; cease to will, detach yourself from life, become disinterested; in so doing you pass from the waking ego to the dreaming ego, which is less tense, but more extended than the other."

We've kept one note to introduce our own solution:

Havelock Ellis, in a lengthy discussion of dreams resulting from gastric disturbances, quotes instances in which the same dinner menu, after a lapse of ten years, resulted in precisely the same dream.

... As, other things being equal, it should! For here is the simple stuff that dreams are made of: awake or asleep, we breathe—inhale, exhale. And with each breath smell-stimuli from without, from within, are brought to the nose—to a smell-sense awake and on the job. As involuntary messages they reach the subconscious part of the brain which, during sleep, is freed of conscious restraint. For the taboo sleeps with the consciousness that causes it—and the smell-messages, without restraint, without co-ordination, wander through the sleepless portions of the mind, and in their wanderings grow great. The symbolism usually attached to dreams seems unnecessary. They are not symbolized so much as they are associated without judgment.

My nose, as I sleep, receives from within me a message of gastric disorder. Conscious judgment being asleep, my subconscious mind may associate this with any odor in my smell-memory that resembles it. And this association may be of small things with great; of grotesque with beautiful. The one and only requisite the Subconscious demands is that the odors be alike.

Next, from outside perhaps, a faint odor comes. The sense of smell reacts to it; transmits news of it to the Subconscious. The Subconscious associates *this* with the first similar odor memory it finds—and adds the new associated, attributed memory to the pattern it has already begun.

Now I am in the midst of an absurd dream—a symbolic dream, if you will—made up simply of one external and one internal odor—with the smell-memories my subconscious mind has furnished to explain them.

As the odor-combinations change with each fresh breath of air, so does my dream progress. The thread of connection, probably, is simply the continuing internal odor which must be linked in, always.

And—where's the mystery?

Dreaming is life reduced to an earlier form—life without motion, vision, judgment, and little sensitiveness to sound, for the deaf and dumb have precise and highly emotional dreams. Children's dreams are simpler than adults'—because children have fewer associated memories out of which to build their sleep-patterns.

Emotions are odor-responses. Dreams are odor-responses. Of course dreams are emotional!

It is interesting that the one dream which Freud can cause in his own sleep as often as he likes is patently an odor-response. He eats anchovies, olives, or other strongly salty food, and goes to sleep. After a time he gets thirsty and awakens. Previously, he always has a dream of drinking.

His nose gets the thirst stimulus. His subconscious mind receives the message—and searches in the archives for similar messages. Every such message in smell-memory is connected with its satisfaction—drinking. And so Freud dreams—of drinking.

Freud and Jung assert that the mythology and symbolism of human culture began in the Unconscious with ancient religious dreams.

From a popularization of psycho-analysis: *What Dreaming Means to You*, by Mary Stewart Cutting, Jr., we cull:

Ancient Religious Dreams:

1. The dream or vision that came with a fast before initiation into a society or cult.
2. The dream caused by incubation—the period of mental brooding before initiation—retiring to sleep in some unusual spot, such as a temple or the top of a mountain, and then awaiting the manifestation of a supernatural revelation.

It seems scarcely necessary to point out that either a fast or a change of environment (a pre-requisite of incubation) must have caused a change in olfactory stimuli received during sleep—and thus a change in dreams.

Fasting, in particular—because it lessons the body-odor component—always results in a heightening of sensitiveness to objective odors.

"Inside a temple." The authoress speaks of the Grecian oracles at Delphi—of the priestess seated on a tripod over a fissure from which dense vapor (caused by the burning of laurel leaves) came, which gradually overcame the young woman until she reached a trance-like state or fell in a fit on the ground.

"The more pious of the seekers were advised to sleep in the temple of the oracle and receive answers to their questions by revelations in dreams and visions *superinduced by the holy atmosphere*. For many hundreds of years, this was believed to be the most solemn of all spiritual communications."

... I leave it to you. Freud—or the sense of smell?

Democritus believed the cause of dreams to be the phantasms or visionary semblances of bodily forms, which floated about in the atmosphere and attacked the soul in sleep.

The old pagan was nearer to the right answer than science has ever been since.

CHAPTER SEVENTEEN

HYPNOTISM AND FREUD

POSSIBLY, in the chemical laboratory that is the mesmerist's body, certain curious substances are formed, and diffused from him as odors. Or, it may be, the man possesses, to a greater degree than most humans, the power of throwing off such odors. However sent, these odors—like certain drugs—act as strange stimuli on the subject's nose—and he is hypnotized. The characteristic movements of the hypnotist's hands, the fixed gaze—seem a means of arresting, or at least causing to function at a constant rate, the chemical change in the visual purple which is sight. In this way, sudden disturbing sight-impulses are prevented from reaching the brain, and the odor messages the subject receives have the utmost possible effect on his mind.

However, since the queer art is no everyday problem, it has little place in this book. But in one former hypnotist who, with his followers, still makes use of actual hypnotic methods, we are deeply, vitally interested. For his entire theory is unconsciously based on the sense of smell; his practice is become a perversion of it.

Freud—and psychoanalysis!

The man's own autobiography, in which he tells how psychoanalysis began, proves it!

He was, he tells us, a student of hypnotism. Grave doubts arose in his mind as to the use of hypnotism even as a means of catharsis. First, he found that even the most brilliant results were liable to be suddenly wiped away if his personal relation with the patient became disturbed. It was true that they became re-established if a reconciliation could be effected, but such an occurrence showed that the personal emotional relation between doctor and patient was after all stronger than the whole cathartic process, and it was precisely that factor which escaped every effort at control.

... *The personal emotional relation between doctor and patient!*

Freud then remembered an experiment he had witnessed with Bernheim. When a hypnotized subject awoke from the state of somnambulism, he seemed to have lost all memory of what had happened while he was in that state. But Bernheim maintained that the memory was present all the same, and if he insisted on the subject remembering, and, at the same time, laid his hand on the subject's forehead, then the forgotten memories used to return in fact, hesitatingly at first, but eventually in a flood and with complete clarity.

... Provided he laid his hand on the subject's forehead!

Freud determined to act in the same way. He abandoned hypnosis, only retaining his practice of requiring the patient to lie upon a sofa while he sat behind him, seeing him, but not seen himself.

... Thus, although he ignored the fact, Freud arranged it so that, of all the patient's special senses, only the sense of smell could be affected!

And, Freud tells us, his expectations were fulfilled.

He was on the right road. If someone had kept pointing out to him —as we have been reminding ourselves over and over again—that knowledge comes to the brain through the senses, *and in no other way,* inevitably he would have reached our conclusions about the sense of smell.

We know that he did not. Yet he moved a little further in the right direction.

"How had it happened," he wondered, that the patient had forgotten so many facts but could nevertheless recollect them if a particular technique was applied?

"Experience—gained from pathologic material—of the frequency and power of impulses, of which we know nothing directly and whose existence has to be inferred like some fact in the outside world—forced psycho-analysts to take the concept of the Unconscious, with no alternatives."

You see, he had four questions to answer: Where? When? Why? and How? Where in man's mind were these lost facts? When did they reach there? Why did they go there? How did they get there?

He answered one of them: Where? They were, he stated, in the Unconscious—the hidden mind. And his answer, because he was reasoning from an observed experiment, was entirely correct.

But on the second question he went inexplicably astray. He was to all intents, remember, repeating Bernheim's experiment on a hypnotized man—and in it the "lost facts" manifestly entered the hidden mind during the subject's preceding period of somnambulism. Despite this, Freud's answer to the question: "When?" was a sweeping: "In infancy—before the child is five years old."

The conclusion was refuted by the very experiment it was based on! But he overlooked that. He overlooked his and Bernheim's experiment entirely ... and his answer to the question: "Why?" was—in capital letters—SEX.

He had to make this fit in with his Unconscious mind and his infant-memory answers. And so he invented his famous Libido, his Oedipus and Electra complexes, his I and ID for conscious and unconscious mind, and his Censor in between.

The fourth question: "How?" (by what sensory means?) he ignored entirely. To this good day he and his followers have ignored it. The Freud-Bernheim experiment still waits for an answer—still *requires* an answer . . . but psycho-analysts today have forgotten their hypnotic beginnings.

Freud had his startling, fascinating theory. And, as a man with a theory *will* do, he set out to prove all things thereby. And since sex enters to an extent into almost every situation, he found his task amazingly easy. All he had to do was to probe until he discovered a sex-angle —and then drag it triumphantly into view as the ultimate cause.

Very quaintly he states: "Psycho-analysts do not say that sex is at the bottom of all neuroses. They simply say that sex is at the bottom of every situation they have encountered."

And since, as he himself says: "The psycho-analyst must make up his mind that the material which the patient lays before him *must be interpreted in a special way*"—we see why this is so.

He is right thus far: A disordered sex-life, a breakdown in the normal functioning of the sex-drive may very well be the cause of a neurosis. The Unconscious may contain its full quota of sex-thoughts. But sex is not the only possible cause of neuroses—and the hidden mind contains other things besides it. And Freud's Libido and Super-Ego and Oedipus Complex are simply big words—psychological words, made-up words —to match the made-up things they describe.

A neurosis, certainly, is a form of cerebral message. As such, it must have a sensory source. Consideration of the several receptors' properties and powers reveals that only one sense can be that source.

And so we achieve these definitions: A neurotic is an individual whose subconscious mind is suffering from the effects of a strange and persistent odor; and his neurosis is the emotional disturbance by which his consciousness is apprised of this suffering. Freud's Censor approximates our taboo, but it is concerned with one thing only: that smell-messages shall not enter the conscious mind unaltered. And, roughly, Freud's I is the conscious mind; his ID the subconscious—and his Libido we return to him unused. The messages which the taboo inhibits are, in adults, of two sorts only: (1) of strangeness or familiarity; and (2) of sex. In children they are of the first type only—for smell-messages of sex have their first beginning with the inception of puberty. There never was a baby who feared castration by his father; or who felt, at two years of age, an incestuous lust for his mother.

Look at the actual process of psycho-analysis! The analyst probes for a sex-urge. The patient, because his conscious and *un*conscious minds tell him the analyst is wrong, indignantly denies any such possession. Little good the denial does him! This very resistance is sublimely considered part of the course of treatment—and termed: "sickness profit."

Throughout, the psycho-analyst remains friendly, helpful, kind. He has plenty of time at his disposal, for (as Freud himself says) psychoanalysis is a matter of months, sometimes of years. The analyst is dealing with an emotionally unstable person—and in this dealing, uses a certain hypnotic technique. The result is that the psychoanalyst, to the subconscious mind of the neurotic, actually becomes a friendly, reassuring, comforting *odor*.

"And so," this is Freud, "the neurotic co-operates with the analyst simply because he believes in him—and he believes in him because he gradually develops a certain sentimental trend toward the Analyst."

Or—legitimately re-stated—"The neurotic co-operates with the analyst because he gradually develops a certain sentimental trend toward the analyst." Freud asserts frankly that this trend is of an amorous nature, disregarding differences in age, sex, and social position. He says that love assumes the place of the neurosis—one disturbance substituted for another—and that this transference, from original repression to love of the analyst, is the "same dynamic factor that the hypnotists have named 'suggestibility'—which is the agent of hypnotic rapport."

But *we* say: the neurosis may be caused by external or internal stimuli. It may be caused by bodily disturbance, or by perfume, or by certain persons or things in the neurotic's environment. It is succeeded, if a cure is effected, by the pleasant, reassuring odor-stimulus of the psychoanalyst . . . and a tapering-off after-treatment, supposedly, ends the patient's need of that.

If there's failure, the analyst has his alibi, for in such cases: "The patient is crazy." Or, as Freud puts it: "When there is no transference, or when it has become entirely negative, as happens in dementia praecox or paranoia, then there is also no possibility of influencing the patient by psychological means."

Psychoanalysis is a queer potpourri of fair fact and dirty fiction; of crystal truth and cloudy thinking; of right practice and wrong theory. But the important thing, the tremendously important thing is that Freud and his followers have learned how to treat and, in some cases at least, cure these neurotics, these sufferers from strange and persistent odors, by suggestibility, transference—*some sort of smell-substitution.*

In criticizing the theory, Dorsey says: "Any psycho-analysis which neglects the facts of genetics and visceral behavior will never discover the materials with which to synthesize a human being. Human psychology is rooted in living human protoplasm and can be explained only in terms of its antecedent history."

Certainly our smell-hypothesis can not be accused of any such neglect. And the technique of psychoanalysis is this hypothesis—at work.

Freud is wrong—utterly wrong—in making the sex-urge unique. We humans aren't nearly so bad as that. We're too close to the beast that takes its sex seriously one season in the year—and lets it alone the rest of the time. We're working as hard as we can to reach the Freudian state—but our very animal nature makes us fall far short of this immoral ideal.

Unknowingly though, and unconsciously, Freud has given us a worthwhile discovery: that some way, somehow, inward devils can be driven out and replaced: one odor substituted for another.

Some day we'll know more about these substitutions—and there'll be fewer insane asylums, as a result.*

---

* Mention might be made here of the mysteriously acting tranquilizing drugs which the psycho-analysts, staring at their suddenly empty couches, view with such disfavor.. By our theory, they are not harmful, unless used to excess, and their action is not mysterious: a felt chemical want is supplied, and a chemically unstable body is made stable again. (1960.)

CHAPTER EIGHTEEN

A SYNTHESIS

BERMAN SAYS SOMEWHERE that a synthesis between the various schools of psychology is latent—that after all, it is a matter of looking at different horizons.

The thought has merit. Each school is observing the same creature —man. Conclusions must be drawn from the same experimental data. Each seemingly opposed doctrine must have in it a germ of the same truth. And so—all the various theories, once the right explanation is found, should fall into it like balls into a slot.

We have seen how our hypothesis has room for the James-Lange theory of emotions. We have found it will take in the entire *practice* of psycho-analysis—though it rejects the perverted theory. We have let Havelock Ellis carry it almost all the way in explaining dreams as messages from the sleepless sense. . . . How about some of the other names and theories in psychology?

Kempf avers: that the behavior of human beings and the higher animals is motivated by the autonomic nervous system and executed by the cerebro-spinal system; that the former system is responsible for the "Why?" of behavior and the latter for the "How?"; and finally, that the cerebro-spinal nervous system is subordinate to the autonomic. "Through its system of exteroceptors, it guides the organism about in the environment in a way physiologically satisfactory to the autonomic system." He regards drives as "states of tension in the autonomic system that demand relief." He practically regards pleasantness and unpleasantness as the determining factors in motivation.

With all this, our hypothesis agrees. It adds, simply, that the inner control for the autonomic system, and the exteroceptor on which the autonomic system depends, are—reading from left to right: the sense of smell.

Berman assumes that temperament is almost exclusively the product of endocrinal activities. He states that there is a temperament for each gland—or for certain glands acting together. Our theory gives importance to two glands only—posterior-pituitary and adrenal medulla— and we assume temperament to be a product of all the bodily metabolism, not of the glands alone.

There's no real conflict . . . merely a difference in emphasis.

By the study of abnormal behavior, other theories have been formed. By them, normal persons are classified on the basis of abnormal groups.

Jung describes two opposed types of personality as introverts and extroverts. He asserts that the introvert lives in a subjective world; the extrovert prefers the objective one.

. . This same separation can be made on the basis of our own theory according as the subjective or objective odors are the more important components of the messages the nose receives—and sends on. And this is the elemental division. In his introversion and extroversion, Jung simply observes the effects of it.

The thing's manifest. Pick out the heartiest extrovert you know. Notice how strangely introverted he becomes when he's sick . . . i.e., when his subjective odor suddenly preponderates.

Freud, in speaking of his erstwhile pupil, says: "Jung attempted to give to the facts of analysis a fresh interpretation of an abstract, impersonal, and non-historical character, and thus hoped to escape the need for recognizing the importance of infantile sexuality and the Oedipus complex, as well as the necessity for any analysis of childhood."

Our theory is neither abstract, impersonal, nor non-historical—yet with it we escape all these necessities.

Adler contends that all emotional complexes can be resolved into a general inferiority complex. His observations lead him to the conclusion that all neurotic patients are physically defective. He believes that the patient's abnormal nervous and mental conditions are the physiological effects of his consciousness of physical inferiority.

If all these defects were organic or visceral, we could easily explain the resultant neuroses, under our hypothesis, by a phrase borrowed from the title of this book. The neurotic is organically imperfect and his nose knows it. Interoceptively, as in ordinary sicknesses, it senses the body's continuing trouble.

Adler, however, lists stammering, stuttering, and so on—of which the nose can know nothing. Therefore, we must seek farther and dig deeper for our connection.

The physical defects as a rule (Adler states) go back to early childhood. . . . A child thus handicapped is forced to realize that he cannot compete successfully with his normal playmates. He is the target for ridicule—and as a result, he becomes profoundly unhappy. He becomes morbidly hypersensitive, distrustful, jealous and envious.

This is enough. The child, in other words, lives in an atmosphere of strangeness. The smell-messages that reach his brain from outside are all unpleasant, unfriendly ones. He is in a continually disturbed emotional state—which must inevitably have its ill-effect upon his body. An emotion-torn body is to a degree a chemically unstable body—and news of this condition is brought to the sense of smell.

The inferiority-complex is an inferiority-odor, at bottom. And—the neuroses resulting from it are smell-responses.

Finally, we look at Behaviorism—"The Religion called Behaviorism," it's been termed. Professor Watson throughout depends on conclusions drawn from animal experiments to explain human behavior. He discards the question of mind entirely, and he calls thought simply subvocal speech.

Let's turn hypercritical immediately: The Gestaltists and independent students of animal behavior have certainly proved that animals think. It would seem that the Behaviorists are faced with the difficulty of explaining *how* animals think—when they have no vocabulary to use in subvocalizing.

The Behaviorists also, Herrick tells us, ignore awareness. I quote from Berman: "sensation, feeling, awareness or reception, it remains a mystery. That mystery is at present impenetrable. That it is not measurable, that it is not directly observable is no more of an argument against it than the parallel question concerning the existence and nature of electro-magnetism that might have been raised in the 13th century."

Of awareness, various unsatisfactory definitions have been attempted. To the psychologist, it comes as a vague, insistent reminder of something forgotten yet alive, present, actually and intimately concerned with whatever he is engaged in, whether inspection or introspection . . . something he should not overlook, something he *is* overlooking.

That there exists this feeling proves that the taboo is no eternal, insuperable barrier. For here is sensation seeping through into its ancient, blocked-off channel. . . . Awareness is the indistinct, involuntary *consciousness* of smell-stimuli!

It is a factor in every experiment made either on animals or on man. If it is disregarded, the experiment is worthless. Behaviorism is chiefly remarkable for the tremendous number of such worthless experiments it has made.

Because the resultant conclusions contain large gaps and broad, empty spaces, the doctrine is noteworthy also for a fervid and unyielding insistence by its devotees that two and two make seven. What our hypothesis does here is to improve the arithmetic by furnishing at least part of the unknown. Two and two—*and* two—almost make seven—human conduct. Behaviorism *plus* the sense of smell should approximate—the answer.

But the older faith must for once clean and order its house, and forever stop disposing of rubbish by sweeping it under the cradle.

Watson's inadequate hypothesis confronted—and continues to confront—him with a mushrooming collection of human actions and human emotions and human memories—and a human subconscious for which he has no explanation at all. Like Freud, he dumps the whole mess into a place from which he is fairly sure there will come no denial —the mind of the very young child. Anything and everything Watson can't explain he says blithely *is* explained by the fact that the child learned it before birth and during the first two years of life—and it is Unverbalized Behavior.

Outside of Adler's insistence on better care of handicapped children, which is logical and right, there is no fire whatever to justify all the psychologic smoke that has been raised about young humans. Many of them, no doubt, are mistreated; are not given the start in life every mother's child is entitled to. But all this talk about character forming— of personality hardening into an iron mold—during the first few years of life is so much poppycock.

Temperament is a matter of bodily chemistry and chemical environment—and change in either will change temperament at any age or time. The child lives faster than the adult; is more responsive. The change will therefore come a very little more swiftly. That is all. It may be that the childish brain can be scarred—grooved. It is also entirely possible for the adult brain to be so affected. Some theorists hold that the infant mind is more plastic—can be scarred more deeply. As to that: any occurrence so poignant that it makes a lasting impression on temperament will be grooved deeply enough to remain in memory, too. It will not be forgotten, and it will not be unverbalized.

On the other hand (remember, we're dealing in new ideas in this book!) it may be that the young child's brain isn't so plastic—easily scarred—after all.

It's a digression from our main subject, but let's consider the possibility.

To start with, here are three seemingly unconnected facts:

(1)   Somewhere between the ages of two and four, permanent memory begins. Before then, everything fades—for reasons unknown.

(2)   The belief is (among those who should know) that individual intelligence—brain power—depends not on the organ's size, but on the labyrinthine intricacy of the brain's surface—its grooves and foldings, infoldings and involutions; that every thought, memory, association leaves its mark in some slight added twist or whorl or grooving; that, so to speak, man's mind wrinkles its way into wisdom. . . . For the argument's sake, at least, we may accept this as true. There seems no reason to doubt its factuality.

(3)   In the race to attain bigness, certain organs outstrip the others. The entire head structure—including our present interest, the brain—grows so swiftly that by the second or third birthday, it is measurably close to adult size.

Putting these facts together, we get: (1) Early memories always disappear; (2) Memories are surface marks or groovings; (3) The brain's swift stage of growth occurs always before, never after, the third birthday.

Now—the only way a three-dimensional mass can grow is: *out*. And in this growing, there's pressure from within, tension on the surface— a push-and-pull that streamlines this surface, stretching out, smoothing out all irregularities. Dry, wrinkled peas, for instance, plump into smoothness as they swell in water. And the crepe-y surface of a much-used toy balloon becomes sleekly round, when inflated. Increased girth in a tree levels out surveyors' marks, lovers' deep-cut initialings.

And similarly also, increase in the size of the infant brain may cause a leveling out of surface inequalities . . . and this means: a smoothing-into-oblivion of the marks and foldings of memory. Exactly the same surface-stretching must occur. It seems at least possible that it may lead to this in-no-way different result.

In other words: *the child's brain may outgrow its memories, precisely as its body outgrows its clothes!*

If this is true, it plays havoc with Unverbalized Behavior and the Oedipus Complex, since it confronts these theories with the impossible task of carving their permanent hieroglyphics upon an impermanent base. To our hypothesis it does no damage whatever. We need only to shift our weights a little, and make the taboo more of an inherited, less of an acquired characteristic.

But not too much less. In any and every case, some part of it must be acquired. The taboo has a battle-of-the-century to fight every time a new child comes into the world. The baby is so willing to forget the interdiction. He has almost forgotten it in his mother's womb. For even thus early, the sense seems at work. Remember: it develops long before the other special senses. As in sleep, it is equipped to function, and in all probability does function—before birth. True, there is no respiration ... but odors may affect the olfactory area through diffusion—the spreading, rising of odorous particles.* In this manner news comes of change (1) in the foetus' own personal odor, (2) in the surrounding fluid, (3) in the all-important maternal blood-stream.

A mother feels life stirring within her, the baby moving, kicking. . . . Random movements? . . . *There aren't any, in life!* Try to think of a single one you've ever made, or ever seen made by any living creature. They're automatic, involuntary, inexplicable (to man's ignorance), unsuccessful often—but never random. There's always a reason, a purpose behind them.

The unborn child's movements are *motor responses!* Responses—to definite sensory messages. And this implies that some part of mind— the oldest part, the only part, at this stage, sufficiently developed to function: the nose-brain—must be at work; implies also that some sense— touch or smell, with no alternatives—must already be reacting to specific stimuli.

The thing's largely conjectural, of course. But whether the nose-brain functions before birth or not, certainly it is the first part of mind to function afterwards. And—it only becomes the Subconscious after birth, when eye and ear come into use and the taboo is re-learned.

The old folks believed in pre-natal influence. Modernists say it's impossible. And it is absurd for a woman to think that by listening to symphony records, she can make her unborn child a musician; or that by looking at great pictures, she can turn him into an artist. . . . And yet—

Until the cord is severed, the blood of the mother is essentially the

---

* Zwaardemaker demonstrated this—in adults.

blood of the unborn child.* Emotion, we know, causes the release of glandular secretions, affecting the blood. Therefore, any emotion felt by the mother must affect this blood she shares in common; must cause —so far as development permits—an identical emotional disturbance in the child. The mother in whose veins adrenalin has been released, carries a child which the adrenalin is affecting. A terror-stricken mother bears within her—unborn, a terror-stricken child.

The thing is manifest. And since, admittedly, too-strong emotion leaves unfortunate after-effects, how much more unfortunate must such after-effects be upon the much more affectable embryo!

We've wandered afar these past few pages, chasing will-o'-the-wisps of conjecture. Quite likely, it was a mistake. . . .

One thing of value we've gleaned, though: In the child, the danger period—the affective period—is before birth. Afterwards, even if— *especially if*—his parents have never heard of Behaviorism or Freud, he is safe—except from the re-establishment of the taboo, and the consequent necessity for the divided mind.

But—this is the real injury, the grave and lifelong harm.

"Every child," Dorsey says, "is entitled in civilization as in savagery, to the full development of its normal inheritance. Civilization has taken curious, often monstrous bents; and even now, in many places, does not hesitate to deny children the free exercise of their human birthright to develop sound minds and sound bodies."

In many places? Everywhere!

---

* This is denied, because there are permeable walls—barriers, supposed filters, between. As to this: the passage may be by seepage, rather than by force-pumping, but there's still a definite in-and-out movement, a truly-linked circuit. And—as to the particular permeability of the substances we're discussing through these walls pass oxygen, waste matter, building materials. The latter *must* include glandular secretions. Surely the fact that these *do* pass through is proof enough that they *can!* Anyway, these permeable walls—common to humans, beasts, and birds—are not such marvellously selective filters as heretofore supposed. One practical experimenter (a poultryman) has lately proved that even foreign coloring matter will pass through them. He has developed some secret way of feeding his hens so that they lay green-yolked eggs.

CHAPTER NINETEEN

PSYCHISM—OLD STYLE

WE HAD BEEN DRIVING our ambulances in convoy, forty feet apart, fifteen miles an hour, all day long. Late in the afternoon we reached our camp—a green spot under a clump of trees. Here were long unpainted sheds, water, wood, and a place to cook.

A company of French soldiers was marching away as we arrived. We sat in our ambulances and watched them pass—little men under longer bayonetted rifles. And we saw that they were not men at all, but boys—younger than we—boys of the Class of 1919, called two years before their time because France so desperately needed them. Sixteen years old they were, and we—who were eighteen at least—felt old and shamed, because we were ambulance drivers and safe, and not real fighting men.

Here, in this camp under the trees, they had received their final training. Now they were leaving it, to go to the front for the first time. That night the trenches were waiting, and we, watching, saw that the lads knew it. Something about their eyes, something about the way they looked around them at checker boarded field and low motherly hill and the little stone village down in the valley—told us so.

But they lifted their rifles gallantly at the word, and they marched away. . . .

"Compagnie a droit! . . . Droit!"

They marched away, singing of Madelon—of Madelon, the so-gallant lady so many gallant men have sung to—down the white road and over the green crown of the hill. And their voices came back to us, softly and more softly through the lengthening shadows:

". . . Quand Madelon
Vient nous servir à boire . . . ."

And so we took over the camp. The largest shed was our barracks. Inside, lengthwise, low platforms, straw-covered, stretched along each wall. On these we laid our blankets. The place was all one room, except that diagonal planking cut off the corners near the entrance, and made two small closets. But the doors of these were closed, and none of us bothered to open them.

Someone turned on the lights. We had just finished making our bedrolls when the French soldier cook appeared in the doorway, ladle in hand, drunk as usual, howling that "La Soupe" was ready.

We raced after him to the dining-shed. There we ate our usual stewed-sheep-and-carrots; talked; sang; laughed—and finally came back to our barracks. It was late—and we were tired—so we voted to turn the lights out, almost immediately.

But then—we found that we could not sleep. Straw rustled; bodies turned; the platforms creaked. Something was wrong. Something was queerly, terribly wrong! And—all of us knew it!

There was a feeling of oppression—of heaviness—of a Presence in the air about us.

One boy rose and began writing a letter by the light of his electric torch. He sat, wrapped in blankets, a ghostly, shapeless figure in the light reflected from the shiny top of his trunk.

"Put that damned light out!" we told him.

"I can't sleep," he protested—but he put it out.

Afterwards, it was no better. We tossed and turned, and the straw beneath us crackled like fairy pistol-shots. . . . There was something wrong—something inexpressibly wrong. And how we knew it!

Some of the boys left to sleep in their ambulances; others took a walk up the hill; the rest of us lay there—wide awake—filled with a strange, boding sense of horror.

And finally, someone turned the lights on again. For no reason at all, I found myself getting up and moving towards the unopened closet to the right of the door. Another boy was there before me—and just as I reached him, he opened the door.

Inside, swinging a little in the air-current set up by the opening of the door, was the body of a French soldier. He had hanged himself to the ceiling with his long, knitted blue waist-muffler. Afraid of death, perhaps—he was only sixteen years old—afraid rather that he could not face death like a man. And so—he had taken his wrong way out.

We cut him down—and we sent word to our commandant—and after a little, men came and took the body away. Afterwards—a long time afterwards—we slept, not very well.

I remember we never slept well in the place. Even the boys who had slept in their ambulances that night—who had never seen the body— said that weeks later something still seemed to be wrong with the room. Personally, I do not know . . . for I had long since moved to my ambulance. We said—those of us who tried to explain it at all—that the boy's ghost was still haunting the place—and we spoke of other haunted houses we had known.

I thought of it many times afterwards. And I have heard, since, of other strange cases like it.

Unquestionably our smell-hypothesis explains the thing fully. It furnishes a simple explanation of haunted houses; of the eery feeling evoked by spots where violent death occurred; of graveyard fears; of the old superstition that ghosts linger around the spots in which, during life, they spent most of their time—for there, it must be, their odor remains longest, and can most strongly evoke the living images out of old memories.

Odors live so very long. Sir Arthur Evans tells of the discovery of the Cupbearer statue in a buried chamber at ruined Knossos. It was not removed immediately from the painted room in which it rested—and an old native was told off to watch the place the night after the discovery.

He fell asleep—but the statue came alive to him in a dream, and he awoke with a start. He was conscious of a mysterious presence; the painted animals on the walls began to neigh and low. There were visions about.

"The whole place spooks!" he said, next morning.

Odors live so very long.

CHAPTER TWENTY

PSYCHISM—NEW STYLE

TELEPATHY AND CLAIRVOYANCE

MY FRIEND MR. X, famed movie director of the silent days, is a believer in the psychic—the mysterious sixth sense. Something happened to him once. He terms it beyond explanation, more than coincidence. This is the story:

Two little boys, X and another, were neighbors in the long ago. And they knew friendship so firm that almost, it seemed, they shared their thoughts together. They were closer than brothers, X says. There was a sort of spiritual kinship between them.

But—when they were ten years old, one family moved to another town, so that the lads lost touch with one another. X grew up; married; had children. Thirty years had passed since he had last seen his friend, and he was forty years old at the time of the following incident.

He was working as an advance man for a circus, traveling through the country in charge of a crew which put up signs and 24-sheets in each place a month or so in advance of the show date. One evening he came to a town of perhaps eight hundred inhabitants; spent the night at the hotel; put his bill-posters to work next morning; and then, since he had never been in the place before, asked his way to the postoffice.

It was just down the street. He went there; got his mail; and stood in the doorway briefly reading it. Then, instead of turning to the right to go back to the hotel and his working crew, he turned to the left. He says he had no reason at all for doing so—except that something seemed to be pulling him that way. He came to the corner, turned to the left again, walked another block and turned to the right. Three blocks further, he turned to the left again. He did not hesitate a moment at any corner—just turned and kept walking as if he knew exactly where he was going. He remembers deciding that the sensation of being drawn in a certain direction was an absurdity; he remembers finally, futilely telling himself he needed the walk.

Just beyond the last corner, he turned again; entered a little grocery store—and found his long-lost, almost forgotten school-chum.

X calls that—telepathy.

If our hero were a dog, there'd be a simpler explanation. And man though he is, the same explanation will serve.

X's friend, a storekeeper in a small town, followed the usual routine of small-town merchants: he went to the postoffice each morning for his mail—and then returned to his place of business. As X stood in the doorway, his nose caught a scent which brought with it old associations —all pleasant—and he followed his nose along the path the author of the scent had followed, earlier in the day.

He had unconsciously trailed his friend!

"Nearly all persons," so Havelock Ellis says, "have possessed, when children, the power of seeing visions in the dark on the curtain of the closed eyelids. I can recall one occasion of its presence, at about the age of seven, when sleeping with a cousin of the same age. We amused ourselves by burying our heads in the pillow and watching a connected series of pictures, which we were alike able to see, each announcing any change in the picture as soon as it took place. This fact—of community of vision—served to impress on my mind the existence of a faculty."

How did it happen that they had mental visions of the same things? Was it another example of minds in psychic tune—of thought transference?

Havelock Ellis couldn't explain it—but our hypothesis seems able to. There was a community of vision—because the same odors in the air—stimuli to the two noses—evoked those visions. The boys were akin; ate together; slept together—and so their own (family, individual) odors were to an extent alike. Their memories, many of them perhaps, were of situations they had passed through in common—and so their associations, odor-brought—were alike . . . Same outside stimulus; same—to a degree—internal odors; same—also to a degree—odor-associations. The resultant—the same mental pictures, smell-evoked.

The whole matter of psychism is disputed, and we are engaged in another and difficult battle of our own. We are simply offering here a practical explanation of mental phenomena Science has not even attempted to explain. Telepathy—mind communication at a distance— may or may not be. At least there are certain analogies in animal behavior.

Standfuss placed a female of the emperor moth, just emerged from her cocoon, in a cage at his window in the heart of town. In six and a half hours he caught one hundred and twenty-seven males—and this in a region where emperor moths were considered rare.

Mell marked males of the Chinese silk-moth, and released them at different points on the railway to seek the females he kept caged. Forty per cent of the males found their way back from a distance of four kilometers. 26.6 per cent found the females from a distance of eleven kilometers—seven miles!

By smell! It is universally accepted that moths use this sense to find their mates.

And the love-lure isn't the only far-felt summons. There's the call of familiar places—the homing instinct. Unwanted cats, abandoned miles away, many times return. Dogs, sold and shipped by train to distant purchasers, often escape and reappear, weeks later perhaps, at their original kennels. Horses, given free rein, bring their lost riders back to camp. The homing pigeon flies, arrow-straight, long miles to nest and young.

Such instances of far-sensing would be termed "telepathy" in humans.

And—it is noteworthy that in most alleged cases of thought-transference, the messages received are of emotional trouble, sickness, or death—things that mean: odor changes.

Tischner, in his TELEPATHY AND CLAIRVOYANCE, has this to say: "Telepathy: On the physical theory, we should have an emission of rays or vibrations going from the brain of the agent, passing through his skull and his skin, through the intervening space, then through the skull and the skin of the percipient to produce impulses and images on his brain. . . .

"But there is no organ of transmission or of perception known, such as we have in speaking or reading, and there is no part of the brain which could be thought to be one. . . . But it seems hardly possible that such complicated results should be obtainable without some special organ. Yet it would be strange to find that a special organ should have been formed for so rare a faculty.

"There is no indication that it is a voluntary transformation and transmission, such as we find in writing and speaking.

"It may be said that it is not a voluntary conversion, but might be an automatic one."

And the paradox is: that he overlooked completely what he has described so squarely: involuntary odor-emission and subconscious smell.

Dr. Tischner accepts telepathy as a fact; finds himself unable to account for it. Ours is the opposite difficulty. There's room for it, but no proof of it, in the hypothesis.

It seems the part of wisdom to remain neutral. Telepathy is no war of ours.

Clairvoyance is, though.

Again we quote Tischner: "The question of clairvoyance is of rather a different nature—as the agent is not a second person, but a thing. Here too the natural scientist is more likely to suppose that rays of some sort pass from the object to the clairvoyant, and act upon him in some way. The problem of how the vibrations reaching the brain are taken up and transferred is not yet solved."

He discusses the possible solutions along this line. Not one of them seems adequate. And eventually he concludes: "We have seen that the ray theories do not explain either telepathy or clairvoyance. The only theory which might yet fit the case is the corpuscular theory. *We should have to imagine that minute particles are emitted by the objects*—say electrons—and that these carry the necessary information."

. . . To the nose!

In his account of psychoscopic or psychometric experiences (the describing of hidden things held in the hand), he finds:

That the successful experiments, generally, not always, are on things which have been in use or possession a long time, such as rings, coat buttons, a sea-shell treasured as a keepsake, a bit of an explosive shell kept as a souvenir . . . Five out of six rings were identified.

He finds also: that the presence of unsympathetic minds spoils the experiments, as the mediums cannot work; that new surroundings and new faces upset the mediums; and that there is a great difference in the rate of success of different mediums.

On one medium, H, he conducted an entire series of experiments. Of one occasion, a failure, he reports: "H said that the things chosen by Dr. Rucknagel (an unsympathetic visitor) had very little OD, and had no action on him."

And, at another time: "H was from the first very much upset by the new surroundings and the new faces. All the experiments were negative."

H describes the OD-messages: "Like visions they seem to come to him from outside; to have an objective character—and he tries to fend them off."

It seems that we can add a syllable and dispel a mystery. For this occult, esoteric OD, the invisible Power immanent in and emanating from objects, becomes, when debunked, simply: odor. And a medium is an individual over whom the taboo, though still present (". . . the messages come; he tries to fend them off,") has not-quite-absolute control.

Most successful experiments, Tischner reports, are on cherished objects, keepsakes, souvenirs . . . Why? . . . Because personal odor clings to treasured things!

A medium is handed a sea-shell wrapped in paper. He holds it to his forehead. The smell of sea-water comes to him, and the odor of sea-life, and the scent of the woman who owned the shell. When he says: he sees a woman by the sea-shore—it's clairvoyance, and a successful experiment.

And—new faces, new surroundings *do* disturb the medium. The presence of unsympathetic minds actually *is* an impediment. For present, too, are strange, unfriendly body-odors.

This last adequately explains the difficulty encountered at every attempted open demonstration of psychism. The results are always negative when sceptics are present. Science, unable to understand why this should be, is prone to shout: "Humbug!" Yet the thing is quite logical. *An experiment on odors can be ruined by uncontrolled odors!* . . . That is all.

Across oceans and continents, migrant birds heed the distant summons of the same old nesting grounds. And eels are drawn from the sea's profound depths back to their parent waters. And spawning salmon return to their native streams.

Man too, like these creatures, may on occasion receive messages unseen, unheard—sensations from afar.

Psychic phenomena! Yet no sixth sense seems needed to explain them—once we give proper credit to the fifth!

CHAPTER TWENTY-ONE

INSANITY

"FIND OUT ABOUT DREAMS," Hughlings Jackson said, "and you will have found out all about insanity." Jastrow points out that not only insanity, but all the forms of delirium—including the drug-intoxications —are variants of dream-consciousness. Moreau of Tours stated that hashish intoxication is insanity—and that insanity is a waking dream. Certain dreams, as narrated by Marro, produce a temporary fit of insanity in an otherwise sane subject.

Havelock Ellis, from whose writings we cull the above, adds that in insane subjects, a dream not uncommonly forms the starting point of a delusion.

We found a solution of the dream-puzzle. The same answer should come close to solving the riddle of insanity. The strange, persistent odors which cause neuroses seem in all likelihood the source of most other and graver mental ills. They may be objective or subjective, of persons, places, things without; or of conditions within. But any overly great, too-long-continued odor-stimulus bodes swift and certain madness.

Temporary insanity gets in the papers most. A murder is committed. The murderer claims he was insane when he did it. There is a trial; alienists are hired, some by the defense, some by the state. They examine the man many times in his cell, and make their solemn report to the court. Three defense psychiatrists avow that he is insane; three state experts swear that he is normal.

You read about it in your tabloid next morning, and wonder if there isn't a law against perjury, and why ordinary liars go to jail while psychiatrists get fees for the practice.

The truth is that any psychiatrist can, with a clear conscience, swear by one book: *pro,* and by another: *con.* He knows that we, all of us, have our psychoses, whether we call them avocations or interests or loves or hobbies (Oh, all right—or smell-theories!), and that the only real test of sanity is whether we're biting the bars of our cage at the moment or not.

Out of a cage, with no bars to bite—as most of us are? That's inconclusive evidence. Inside the cage, not biting the bars? The psychopath, for the moment, is as sane as the next man.

"He was my partner; he robbed me of my business and my wife. Then he came into my house, where I sat alone, brooding—and laughed at me. A sudden fit of madness came over me, and I shot him." This is the murderer's story. Under present conditions, the psychiatrists' tests of the man's sanity are foredoomed to failure—*because the tests are made in the wrong place.*

Inside the steel-barred cell, the air is impregnated with the prison odor, redolent of the auras of other prisoners and guards nearby. There is the lingering memory of former tenants; there is, during the alienists' visits, their personal odor.

But not one of these has any detonating effect on the murderer's sanity. His sudden madness came on him in his own home, where he sat in his bitter desolation, surrounded by the scented memories of the wife he had lost, and it came at the moment when these emotional stimuli were heightened by the odor of the man he hated most—the man he killed.

Perhaps the same concomitancy of odors might never come again, and he could be safely turned loose among his fellows. Perhaps a similar odor-stimulus might come from entirely different components and cause another murder-order to be sent from the association-regions of his brain—driving him, this time, to kill an innocent man. It's hard to tell. But the only true test of latent madness would occur in the murderer's own home, in the familiar atmosphere of the murder-scene, and it would be made conclusive by the sudden admission of the slain man's odor when alive, obtained possibly from his clothes. And the psychiatrist would guard against any possible disturbance from his own personality, by watching the murderer's reactions from a distance.

Television should revolutionize psychology.

Because of the same reason: odor-persistence and its effects—insane asylums seem almost as basically wrong as hospitals. It is difficult to see how cures can ever be obtained in such disturbing surroundings. And this is all the more pitiable, in that the right environment may be the sole remedy needed.

For madnesses produced by odors can be dispelled by odors. Change of scene, of habits, of diet, of associates—one of these may work a seemingly miraculous cure, if the harmful stimulus be external. And even when the troubling, persistent odor is an internal one, the case is not hopeless. Freud has unknowingly proved the feasibility of benign odor-substitution. Osmiotherapy remains to be explored.

Change in any one component changes the whole, of course. Any continued variation, plus or minus, from the body-norm must produce an abnormal cerebral smell-message. There are indications that chemical deficiencies in the constitution of the brain or nerves are definitely linked with insanity, and (this should be interesting!) there are indications also that the deficiencies may be remedied by supplying these chemicals directly to the sense-organ of smell.

From an address before the American Medical Association we quote: "Sensational advances in relieving insanity by use of carbon dioxide and oxygen gases and amytol have recently been made by Dr. Walter Freeman and his colleague, Dr. Karl H. Langenstrasse. 'The effects of known deficiencies in oxygenation of the brain are often striking,' Dr. Freeman said. 'These mental phenomena can often be banished abruptly by supplying an abundance of oxygen to the brain.

'Whether there is a dietary deficiency or whether some toxic substances are formed in the intestinal tract or whether emotionally restricted respiratory exchange leads to suboxidation, it is impossible to say. Nevertheless it is enough to work with and observe that which we are powerless as yet to interrupt.' "

In the minutes of the American Chemical Society we find: "Insanity, particularly dementia praecox and dual personality, due in considerable degree to endocrine disorders was spoken of by R. G. Hoskins, M.D., of Harvard Medical School. These forms of insanity, he said, comprise about 20% of the total hospital population of the United States. In Harvard experiments, a hormone from the thyroid gland was given to eighteen dementia praecox sufferers. Five of these became well enough to go home. Most of the others showed significant improvement.

"Jean Paul Pratt, M.D., of the Henry Ford Hospital, Detroit, said that the clear cut, striking results of experiments on animals are not reproduced in human subjects up to the present time. A minor human difficulty is discomfort from hormone hypodermics. In cases of one hormone, this discomfort was avoided by putting it in a *nasal spray so that it was absorbed by the mucous membrane of the nose.*"

And Havelock Ellis states: "It has been found, by Meunier, especially easy to apply olfactory stimuli during sleep and so improve the emotional tone. In hysterical subjects, he found that essence of geranium provoked various agreeable dreams, followed by a pleasant emotional tone during the following day."

There is a whole world of promise in mental treatments by means of the sense of smell. There are various incenses and odors—ancient, medieval, Arabic, Chinese—which have been forgotten by the medical profession—and which are well worth while discovering again.

Somebody ought to do it.

CHAPTER TWENTY-TWO

THE MYSTERY OF MIND

OUR DEFINITION of the Subconscious is: the ancient, animal brain, still functioning and, despite a strong taboo, still under smell-dominance. In many and diverse ways, throughout the preceding pages, we have tested this concept. We have found no flaw in it, whatever. Put into practice, it *works*.

If we could say that this old brain plus our newer Consciousness represents the sum-total of Mind, we could rest here. But there are other factors—even, it may be, a third grand division. There seems need to discuss these. And so, somewhat dubiously, we attempt to resolve the human mind into its diverse parts:

Memory. Suppose I am faced with a problem. Late at night, just before I fall asleep, I think about it. While I am asleep, my subconscious mind labors over the question. Before I awaken, the answer is found —and filed. I wake up; think of my problem; discover, in my memory, the answer, ready.

Memory is a bin, a storehouse which both parts of the mind can use. We cannot say it is exclusively of the Subconscious or of the Conscious. Much of it is community property—though the joint owners never meet.

Consciousness. Imagine this sequence: Something frightens me; I run; I slow down; I stop running. How much conscious control did I have of the situation?

None! The situation controlled me. For here, I think, is the way my mind worked during the incident: vaguely, my eyes saw something. I turned, suddenly attentive. Subconsciously my nose appraised the object. A fear-message was sent to my brain. Adrenalin was released.

And now, and only now, do I have a conscious thought.

Consciousness: "I am afraid."

No part of my body pays any attention. My legs start rhythmically moving me away from danger.

Consciousness (somewhat surprised): "I am running."

The fear-scent fades with distance. Finally, when my nose decides it's safe, I stop.

Consciousness (with relief): "It seems—I'm safe here," (happily): "It seems—I'm all right!"

The supposed master-mind is singularly passive in stormy times. It does nothing, actually, but discover and report—apparently to itself—the course of events. It's like the broadcaster, not the quarterback, at a football game.

Judgment has been attributed to Consciousness. As regards bodily welfare, this may well be doubted. So long as he stays out of automobiles, a drunk man, entirely without conscious judgment, usually fares as well as his sober brother.

If the judgment be considered as of problems, again we find that consciousness seems entitled to but little credit. All the conscious mind actually does, it appears, is to repeat the problem over and over in a Watsonian, sub-vocal manner until some other part of the brain supplies the answer.

The creative instinct is attributed to Consciousness. I speak sensibly and with no thought of modesty or lack of it when I say that I possess this instinct to some small degree. I have written plays and seen them produced on Broadway; I am a practicing screen and radio writer; I have invented machines and patented them; I have evolved this new theory of ours. And—I am positive that consciousness has had nothing to do with any of the above forms of creation. In writing, my conscious mind reads what I write *after* I have written it, and reads it exactly as it would read writing by another hand. When I was faced with a problem of mechanics, when some gadget I was working on wouldn't work, the answer—if and when I found it—always came to my conscious mind in the form of impatient speech, apparently addressed to consciousness: "Fool! don't you see that *this* is wrong?"

Consciousness, I believe, is a chronicler of events, a repeater of questions—and that is all. As such, it is completely Watson's subvocal speech.

The *Subconscious* has powers of association, of judgment, of reason. It receives messages internally and externally, collates them—and acts upon them. It is the mind that is tabooed, but it is also the mind of instincts, of emotions, of personality, and of dreams.

In babies and in animals it is one and indivisible with Consciousness. In young children, it is side by side with Consciousness. Corley, in the *Psychological Review,* speaks of the early use of self-words by a child. He finds that the child distinguishes between itself as (1) body; and as (2), self-assertion united with action. It refers to the former as "Baby," and to the latter as "I."

Side by side, you see, "Baby" is controlled by the smell-autonomic system; "I" is the part under visual-auditory, cerebro-spinal dominance. "Baby" is the Subconscious; "I" the conscious mind.

Conscious, Subconscious—these two—but there is something else, something that asks the questions Consciousness answers as to what the Subconscious is doing; something that solves the problems Consciousness gets credit for solving. Animals do not possess it; humans do. It dictates what Consciousness writes; it seems to be more closely linked to the Conscious than to the Subconscious, but it exercises control over both; it seems to sleep when the body sleeps—and its answers are somehow different from the ones the Subconscious furnishes during sleep.

It is the third part of the mind, I think. And thus—I seem to have rediscovered Psyche—the Soul.

I know less than nothing about it—but it's there. It's not the Conscious; it's not the Subconscious; it's more of a Super-Conscious. And it is apart from and beyond our hypothesis.

With that understood, we can consider what is left—and understand it. We leave Consciousness with its secretarial work—and we repeat that what remains: the mind of instincts, of emotions, of personality and temperament and moods, of love and friendship, of day-dreams and dreams, and—when it breaks down—the mind of neuroses and psychoses, is the Subconscious—the mind of the sense of smell.

CHAPTER TWENTY-THREE

CONCLUSION

ONE BY ONE, we have re-surveyed the ill-mapped bypaths of human behavior. Now, in this last chapter, we check over, briefly, the base-lines we have drawn.

Where did we begin? With the statement that all living organisms are odorous. Surely, this is implicitly so. And then—we said that all organisms (including the amphibians, who have an understandable reason for having partially lost it), possess a sense of smell. Here we departed from the usual dogma—because science has denied this power to man and to the higher apes.

We looked at man. Against our assertion we found only an unreasoning insistence that man has lost this faculty. We examined the organ itself—and found it fully equipped to be a functioning sense. We found that there was evidence to spare of human beings employing this sense almost as usefully as the animals that depend on odor most. We also found that when tested objectively, there is no measurable difference between the olfactory acuity of civilized man and his keen-scented savage brother. We found finally that the sense of taste to a great extent is really that of smell, and that there is no evidence of any loss of power in it. From all these things we arrived at the first statement of our hypothesis: that man has a functioning sense of smell.

But man also possesses an individual, personal, variable odor, and by the very nature of things this must be a component of any stimulus reaching the nose. Every outside odor must blend with it; be altered by it.

Admitting this factor, we completed our first statement thus: *Man has a functioning sense of smell, acted upon by combinations of objective and subjective odors.*

There is a mountain of evidence to support this. No flaw that we've been able to discover weakens it.

If anything in a fallible world is true, this is!

We took up next the immediate question: if man has this potentially useful sense, why doesn't he use it?

Our answer was: he *does* use it! He has no choice—but to use it. By nature and by structure, the smell-sense is automatic. It possesses no shut-off valves or curtains or gates. With every indrawn and outdrawn breath, from birth to death, it must function, know sensation, transmit messages to the brain.

Messages that *must be received!* Scant few of them gain entry into consciousness. They must, therefore, be shunted into the part of the brain that is not conscious: the Subconscious.

We knew, of course, that this side-tracking of sensation was un-natural. We found the explanation in a full-strength, active taboo. Other observers had already (but too casually)* noted the existence of this. We pioneered solely in pointing out its valid implications.

Out of conflict and confusion at the dawn of language (we thought) it arose; through the years it has made men prisoners of race-hatreds and superstitions and animistic beliefs; today it remains a psychic, shadowy spite-fence, and the barrier of it forms the divided mind.

But what about the exceptions, the familiar odors which reach con-sciousness? One type, we learned, could be eliminated at once. The irritants, which affect the trigeminal nerve solely, are no concern of olfaction. We looked at the two remaining groups: food smells, sweet fragrances—and we found, in each case, specific reason for an abeyant taboo. Their presence established the metes-and-bounds of interdic-tion, but in no way denied the basic fact of it.

And this basic fact buttressed our second statement: *Smell messages, consciously tabooed, are subconsciously accepted.*

We know that this is true. We believe—we proved it.

In dreams and sudden friendships and quick hatreds, in moods and manias and many another aspect, good and bad, of the hidden mind we noted the effects of these subconsciously accepted messages. In their individual, progressive affirmation we found collective proof of our third statement: *The total of these acceptances is the Subconscious.*

Inevitably, we realized, this definition would be attacked. It was too new, too revolutionary, too simple.

Yet the concept is physiologically probable; it encompasses every known attribute of the mind beneath; it isn't shot through with the absurdities and exceptions and involvements that beset other, earlier theories.

---

* Cf. Philip Curtiss': "Smells are sensory swear-words."

By adducing it, we even achieved the magicry of reconcilement among these older, warring theories—a synthesis that in itself became a sort of corroboration. For it was as if we had supplied the missing piece in a puzzle; had made a complete picture out of apparently unconnected dabs of jigsaw work. . . . And no bit of falsity could have accomplished this.

By every available touchstone, the third statement of our hypothesis tests true.

Our last finding had to do with man's ancient body, still functioning, still under nose-brain control. Here, and definitely, it seemed to us: what was, *is;* the usual accrediting of the vital life-processes to cerebro-spinal dominance is wrong; they remain as, in pre-vertebrate days admittedly, they were; as, again admittedly, in all creatures below the primates (except the amphibia*) they still are: the business of the sympathetic system and the first of the special senses.

The *exclusive* business, apparently. We struck a trial balance in our fourth statement: *The automatic actions of the body are smell-responses.*

There is (we found) affirmative evidence without end, almost, to substantiate this. More than any other concept, it accords with physiological fact in using known organs instead of imaginary ones. It corroborates, and is itself borne out by, the James-Lange theory of emotions. It works with Lloyd Morgan's Canon of Animal Behavior to make instincts understandable. By the ultimate test: the Law of Parsimony, it is simple, straightforward, and to the last decimal point, almost, adequate.

Yet the rule misses universality. There's that infuriating (to us, because we should be able to explain it) matter of sleep. There are reflexes of a primitively local nature. There are, finally, certain emotional responses: to music; to radio; to the printed word; to pictures—moving, and still—which must without question be based upon eye- or ear- sensation.

These responses are always, and significantly, weaker than the biologically more natural olfactory ones. Poignancy is absent from them;

---

* Even for these supposedly anosmic creatures, and despite outward atrophy of the sense-organ, the rule should hold. The nose-brain itself, so far as I can discover, remains unaffected, able to receive back-door ganglionic and blood messages—and these would suffice.

their quality of vicariousness is so marked as to justify our describing them as: substituted emotions.

Through the ages, probably, there has occurred a sort of cerebral, functional endosmosis, by means of which the "higher" brain centers have penetrated, a very little, into the ancient sensorium so that now eye and ear control that miniscule part of emotional life implicit in the esoteric thrill of a symphony, the publicized horror of Orson Welles, the thrill of Clark Gable's pictured smile. But—no more!

Yet these things *are* emotions; they do not fit the pattern we laid down; we must take them into account.

Throughout this book we have coupled theory with facts. Only in these exceptions to the fourth statement have we come upon contradiction in fact to any part of our hypothesis. Since this single objection is essentially a boundary-line dispute, a reduction in the scope claimed— from universality to predominance simply—obviates it.

We make this small revision.

Amended thus, our hypothesis bears the hallmark of truth. Fully and fairly, it merits addition to natural science. Medicine can use it . . . psychiatry . . . psychology, also. It opens new fields of amazing promise for future research; it offers amazing future hope for us all. For— as our study of the deaf-blind and of young children revealed—the taboo is only weakly inherited, but strongly—by each generation—relearned.

It can be done away with, almost, within a generation.

By ending this taboo, we make for our own happiness, for a greater consciousness of our own bodies, for control over unpleasant emotions, for a heightening at will of pleasant ones. We can treat the sick as they should be treated; we can cure insanities, perhaps; we can get back some share of the joy in life every child is born with.

Man has one power that is almost God-like. What he knows, he can in time master. This book is but first schooling, primer-knowledge— but through it we shall achieve, some day, a new and strange sensemastery.

That day, the spirit will triumph, ending the rule of the flesh. We shall enslave to good purpose our bodies, even as we emancipate ourselves.

We shall learn once more to laugh as little children laugh. We shall end all sadnesses quickly, in little-children-fashion. We shall be done with bitterness; know less of anger, and we shall banish fear.

Here, once again, is the hypothesis—guide-post to all these things:

(1) *Man has a functioning sense of smell, acted upon by combinations of objective and subjective odors.*

(2) *Smell messages, consciously tabooed, are subconsciously accepted.*

(3) *The total of these acceptances is the Subconscious.*

And, primarily and principally:

(4) *The automatic actions of the body are smell-responses.*

Simple ... straightforward ... adequate!

Hypothesis? ...

We mean: the Law!

The End.

ABOUT THE AUTHOR

In 1908, young Jimmy Millen (age 10), rode horseback alone (except for the equally unhappy horse) 150 miles cross-country, through mud and flood, in three rain-swept days. At eleven he was memorizing dictionaries and encyclopedias and straw-bossing 100 field-hands, all of whom knew more about farming than he did. At sixteen, he was the youngest student at Yale. At eighteen, he was a veteran of the French Army. At nineteen, when W.W.I ended, he was a U.S. Flying Cadet.

He returned to Yale to complete his degree. After graduation, he drove a memorable Mercer twelve thousand miles through most of the then roadless Eastern and Midwestern States and Canada—during which adventure he was held up in Massachusetts, wrecked in Maine, arrested in New Hampshire as a bank-robber (he wasn't!), jailed in Kentucky as being the draft evader Grover C. Bergdoll (again he wasn't!), and half-drowned in a Tennessee river flood.

Eight years running a Delta plantation followed. Out of sheer boredom he tried his hand at invention, evolving such diverse things as a self-cleaning rake, a child's cereal dish and a power cultivator. He also began writing.

In 1930 he came to Hollywood. During the next ten years, he wrote ten plays and sold six of them (one to Jed Harris); had two (tragedies) produced on Broadway, one (comedy) put on variously throughout the country. He had screen credit, a dozen studio contracts, a radio serial. Also—he worked endlessly on his major interest, the Sense of Smell.

In 1932 he handed a Mss, of YOUR NOSE KNOWS to his agents. Publishers rejected it; he was after all, a known playwright and screenwriter—not a credentialled scientist. And, theorizing about an active sense of smell in man—sniffing, so to speak, one another's odors, was not at all seemly conduct.

In 1937, the editor of one major publishing house asked to see it again; said he'd been thinking about it for five years. The author dug it out; decided it needed polishing. Once more, he let everything slide, ignoring studio calls, his agents' frantic pleas. Months later, when he finally sent it in, he learned that the editor who'd wanted it had died. His successor wasn't interested. In disgust, Mr. Millen shelved the book. He had new worries: His health.

Mr. Millen surrounded himself with gardens; became a horticulturist for a time. Still, his belief in the importance of his book bedeviled him. It ought to be published, he thought; it **had** to be published. And so it was.

James Knox Millen died March 24, 1988. More than 200 universities, colleges, research groups, government agencies, book stores, and libraries throughout the world (and hundreds more individuals), purchased the book following its publication in 1960.

It was Millen who theorized that "the mind of instincts, of emotions, of personality and temperament and moods, of love and friendship, of day-dreams and dreams, and—when it breaks down—the mind of neuroses and psychoses, is the Subconscious—the mind of the sense of smell".

It was Millen, the Broadway playwright/screenwriter turned pioneer theorist, who said: "Man has a functioning sense of smell acted upon by a combination of subjective and objective odors; smell messages which are consciously rejected are unconsciously accepted; smell has memory." He hypothesized that odors are medicines and cited early experiments whereby typhoid bacillus was killed in 45 seconds by vapor of cinnamon. He identified that each man has an individual odor and speculated that man's sympathetic nervous system, through the posterior pituitary and adrenal medulla down to the very ganglia connecting the brain and coccyx is controlled by nose/smell messages.

YOUR NOSE KNOWS is not a scientific tome. It is a fascinating read; argumentative and marked throughout by the personality of the author—who believed that a thinking man had a responsibility, if he could, to contribute to the general body of knowledge.

Aromatherapy and man's sense of smell are big deals now. This book tells you why.

www.ingramcontent.com/pod-product-compliance
Lightning Source LLC
Chambersburg PA
CBHW020916290526
45784CB00002BA/579

* 9 7 8 0 5 9 5 0 1 2 0 8 4 *